# Pay Dirt

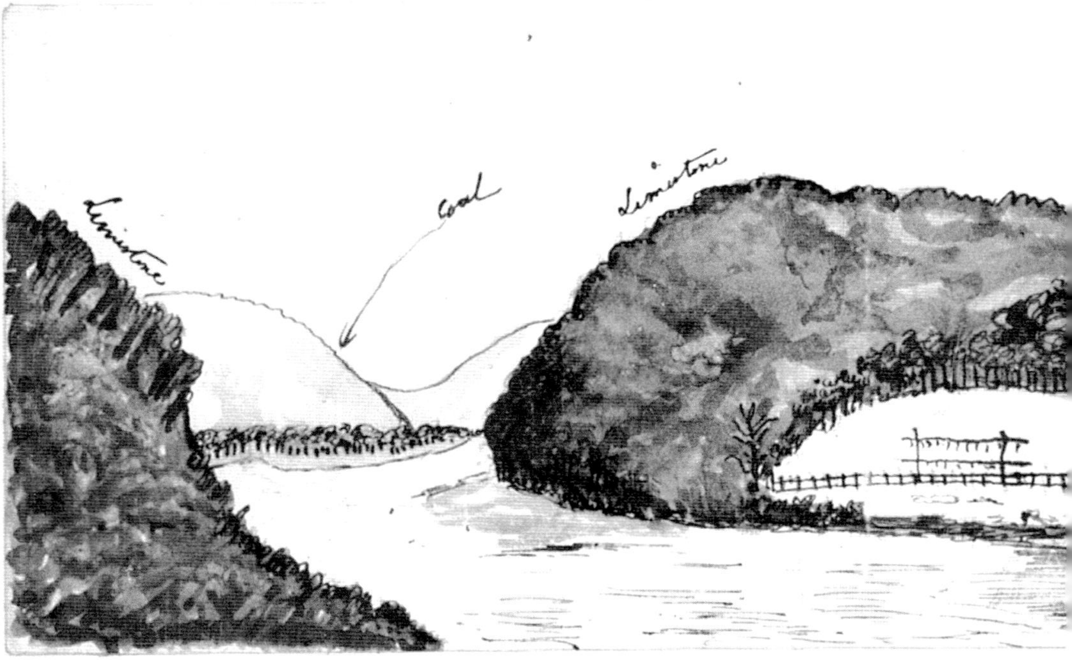
Limestone
Coal
Limestone

# Pay Dirt

## 'The Westland Goldfields', from the diary of William Smart

**Hilary Low**

UP
CANTERBURY UNIVERSITY PRESS

UC
UNIVERSITY OF CANTERBURY
*Te Whare Wānanga o Waitaha*
CHRISTCHURCH NEW ZEALAND

First published in 2016 by
CANTERBURY UNIVERSITY PRESS
University of Canterbury
Private Bag 4800, Christchurch
NEW ZEALAND
www.cup.canterbury.ac.nz



ISBN 978-1-927145-75-3

A catalogue record for this book is available from the National Library of New Zealand.



Front cover image: Drawing by William Smart from 'The Westland Goldfields', p. 89
Back cover images, top to bottom: Drawing by William Smart from 'The Westland Goldfields', p. 85; William Smart (C. H. Manning photograph, Canterbury Museum, 19XX.2.5244); the first page of Smart's manuscript, 'The Westland Goldfields', p. 80
Frontispiece/title page: Drawing by William Smart from 'The Westland Goldfields', p. 101

Book design and layout: Quentin Wilson: quentin@quentinwilson.nz
Front cover design: Hilary Low

Printed in China through Asia Pacific Offset

Published with assistance from the Canterbury History Foundation Inc.

# Contents

Dedicated to the memory of
William Smart 1830–1900

C. H. Manning photograph, Canterbury Museum, 19XX.2.5244

# Acknowledgements

Many people have contributed to this book with advice, guidance, specialist knowledge and services, support, energy, time and welcome diversions. I am indebted to them all.

First, a thousand thanks (not nearly enough) to Ali Carew, my editorial guide and mentor – and sister. I am most grateful to Catherine Montgomery and Katrina Rainey at Canterbury University Press for their guidance, support, patience and good humour throughout the publishing process; and to Canterbury University Press's former publisher, Rachel Scott, for first agreeing to take this project on. My thanks, too, to the book's copy editor, Gillian Tewsley.

I am grateful to the Macmillan Brown Library at the University of Canterbury, which holds William Smart's manuscript, for granting permission to transcribe and publish the work and to reproduce Smart's drawings.

Much of my research for the book was conducted at the Christchurch office of Archives New Zealand. I wish to thank the staff there, and also at the Wellington office, for their exceptional service. My thanks go to staff at the National Library of New Zealand in Wellington, particularly the Alexander Turnbull Library; to Julia Bradshaw, director of the Hokitika Museum; Nicolas Boigelot and other staff at the Documentary Research Centre, Canterbury Museum; Helen Pannett of the Research Facility at Nelson Provincial Museum; staff at the Aotearoa New Zealand Research Centre, Christchurch City Libraries; Margaret Mort and Karen Prendergast at History House Museum in Greymouth; Jackie Gurden, chair of Heritage West Coast; Elaine Marland, information services advisor at Heritage New Zealand Pouhere Taonga in Wellington; and Jane Pidgeon of Heritage New Zealand's Southern Regional Office.

For their expertise and advice so generously given, I wish to thank art historian

Jane Clendon of Mahina Bay, historians Hilary and John Mitchell of Nelson, writers Mark Pickering and Wayne Martin, both of Christchurch, Dan Moloney of Westport and Brian Wood of Hokitika.

My warmest thanks to writer Nic Low for his enthusiasm and support for this project, to designer Anna Woolhouse for her expertise in working with the digital images, and to Jon Terry for reading the manuscript and offering valuable suggestions.

I am grateful to Martin Smith of New Zealand Micrographic Services Ltd for managing the digitising of William Smart's drawings from the microfilm, and to Jo Condon, manager of the Macmillan Brown Library, for her kind co-operation in making the microfilm available for digitising.

I join Canterbury University Press in acknowledging the Canterbury History Foundation for their generous financial support for this project.

On my frequent research trips to Christchurch I enjoyed the kind hospitality of my brother Geoff Low and sister-in-law Hikatea Bull, and I am most grateful for this.

Lastly, my love and appreciation to my young ones, Anna, Jon, Lucy, Nigel, Luka and Eli, and to members of my extended family, for their patience and encouragement during the evolution of this book.

Hilary Low
Wellington
2016

# Introduction

On the West Coast of the South Island, driving along Highway 73, there is a point between Kumara and Jacksons where a traveller may catch a view of two impressive, bush-clad mountains on the opposite bank of the Taramakau River. The peak on the left is Mt Smart. It was named after a man who is now all but forgotten in New Zealand's history. But behind the name there is a story, set in the years leading up to the West Coast gold rushes, about the search for payable gold and the promise of a grand reward.

In many ways William Smart was an unremarkable man. The son of a wealthy London merchant, he emigrated to New Zealand in 1850 as a young man of independent means and became one of Canterbury's earliest settlers. By 1854 he had set up in business and owned properties in Christchurch and near Kaiapoi, and appeared destined for a relatively comfortable and uneventful life in the young colony.

But William Smart became obsessed with gold, and it was this obsession that led him in 1862 to abandon his life in the settled east of the island and spend the next seven or eight years on the rugged West Coast in search of the precious metal. Then, in 1887, long after he had quit the diggings and the gold rushes had declined, he read a statement in a Christchurch newspaper that nobody had yet produced a comprehensive history of the earliest gold discoveries on the Coast. Provoked, Smart sat down and wrote his own history, 'The Westland Goldfields', based on the diaries he had kept during those eventful years. For he had been there from the beginning, and in the very thick of it.

In taking up his pen Smart had a very personal agenda – to set the record straight on the part he had played in discovering a payable goldfield in West Canterbury, on the West Coast. Smart believed he and his mate Michael French were the rightful recipients of the reward offered by the Canterbury government for the discovery of a goldfield in the province. Initially it was a rich prize of £1000, equivalent in purchasing power to over NZ$120,000 in today's money.[1] His resentment when the prize was awarded to another prospector never faded, and became his prime motivation for writing 'The Westland Goldfields'. He finished it in 1890, and died 10 years later without having arranged for its publication; whether he ever intended to is not known.

I discovered Smart's writing while I was researching my first book, *Pushing His Luck: Report of the expedition and death of Henry Whitcombe, by Jakob Lauper*.[2] I had never heard of 'The Westland Goldfields' or its author, but I tracked it down in the Macmillan Brown Library at the University of Canterbury. On first reading, Smart's story jumped off the page. I was equally intrigued to discover his pen-and-wash drawings that accompanied the text.

Some time after it was deposited in the Macmillan Brown Library, Smart's original manuscript unfortunately went missing. It exists now only on microfilm. Access is limited as it may be consulted only in the Library. Furthermore, parts of the text are badly faded and in many places it is almost illegible. Until now it has been referred to or quoted by some other writers on the gold rushes, and some of Smart's drawings have been reproduced, but the whole text with its illustrations has never been published and is therefore virtually unknown.

For *Pushing His Luck* I needed to use extracts from 'The Westland Goldfields', so I transcribed the relevant passages, and also selected a few of Smart's drawings to include. Thinking that this manuscript deserved a wider audience, I went on to transcribe the whole document for eventual publication.[3] I put the project aside for a few years, but Smart and his story stayed at the back of my mind. Finally, I sent a proposal to Canterbury University Press, who agreed to take it on.

What distinguishes Smart's story from most other contemporary accounts of the West Coast gold rushes is that it is based mainly on the diary he kept while on the trail and at the diggings, and it is set mostly in the years before the gold rushes that began in 1865, when only a handful of people, Māori or Pākehā (Europeans), were on the Coast.

But 'The Westland Goldfields' is more than simply a series of day-book entries, or a compilation of evidence to prove a point. Smart was a keen observer of his surroundings; he had a real interest in geography, some knowledge of geology, and an artist's eye for the landscape. He revealed the harsh conditions that dominated daily life on the Coast: the remoteness and isolation, the hazardous terrain and extreme weather, and the very real risk of drowning or starvation. He was also curious about Māori history and practices, and stories about the land. He described the happenings at the pā and on the diggings, as well as wider events surrounding the discovery of gold. Aware of the political machinations at work, he knew that decisions being made in Christchurch would affect his own life and prospects, as well as developments on the Coast. Nor was he shy about writing to men of influence, such as politicians, government officials and newspaper editors, to further his interests.

Many of the men Smart encountered on the Coast were the main players of the day. Through his eyes we see Reuben Waite, William Revell, Charlton Howitt, Albert Hunt and, notably, the chief Tarapuhi, his nephew Ihaia Tainui, and the Poutini Ngāi Tahu people at Māwhera pā, with whom he developed a close and lasting bond.

All those he met were of widely diverse backgrounds, but what mattered to Smart was their ability to survive on the Coast – their resilience, determination and resourcefulness – and their dealings with their fellows. He recorded the fierce competition that sometimes existed on the diggings – the deceptions, violence and even murder – yet also the co-operation, generosity and camaraderie, not only among mates but between strangers and between races.

All this was incidental to Smart's purpose, however, which was to set out the facts surrounding the discovery of payable gold in West Canterbury. He believed he was presenting a true and comprehensive history, but this would have been impossible at the time: communication on the West Coast was limited, and rumour, hearsay and fabrications were rife; and with self-interest fanned by gold fever, no eyewitness account could ever claim to be unbiased. And Smart makes it all too clear that underlying his 'history' was a deep sense of injustice.

Smart may not have been a true historian, but he was a first-rate storyteller. His daily record offers lively descriptions of dramatic events and notable characters, and he had a sharp eye for quirky detail. Nor was he a trained artist, although his apprenticeship in a London architecture and building firm would have given him training as a draftsman. His illustrations are in the tradition of topographical drawing. They are unadorned, strong and honest, showing clarity of line and solidity of form. Almost all of them were drawn in the field, then cut from his original diary and pasted into his manuscript. They are also rare. Some early surveyors, explorers and travelling artists made visual records of the pre-gold rush West Coast, but few works by gold prospectors living and working on the diggings have been uncovered. Several more of Smart's watercolours have turned up in the Alexander Turnbull Library.

Smart was clearly an intelligent, educated man who was driven by his sense of grievance to prove a point. In doing so, he provides us with a close and vivid description of life on the diggings during those early pioneering days on the West Coast, before it was overwhelmed by tens of thousands of miners and camp followers all eager to get their hands on the gold.

The book is presented in three parts. Part 1, 'In search of that golden land', traces Smart's life to 1862, and sets the scene of life in Canterbury during its first 15 years of settlement, concentrating on the search for gold in the province.

Part 2 is the transcription of 'The Westland Goldfields'. While most of Smart's account is closely based on his diary and concentrates on the pre-gold rush years (1862–64), its final section is more of a general memoir of the period from the late 1860s, after he moved back to Christchurch. Episodic in style, it describes some of his return visits to the Coast and the rapid changes he observed as the gold rushes reached their peak and gradually declined. All 18 drawings from his manuscript are included in Part 2.

Part 3, 'A smouldering resentment: Smart and the gold reward', fills some of the gaps in Smart's account, scrutinises the evidence, and tells the wider story surrounding the granting of Canterbury's gold reward. It describes Smart's life and work after he returned from the diggings, and how he came to write 'The Westland Goldfields'.

## Notes

1 This excludes assets, such as houses. Inflation Calculator, Reserve Bank of New Zealand, rbnz.com.

2 Hilary Low, *Pushing His Luck: Report of the expedition and death of Henry Whitcombe, by Jakob Lauper*, Canterbury University Press, Christchurch, 2010. Set in Canterbury in 1863, it records the ill-fated expedition of surveyor Whitcombe who, with his assistant Lauper, was sent from Christchurch to find a pass over the Main Divide suitable for a road to West Canterbury in anticipation of a gold rush there.

3 A transcript of 'The Westland Goldfields' has been deposited in the Macmillan Brown Library at the University of Canterbury.

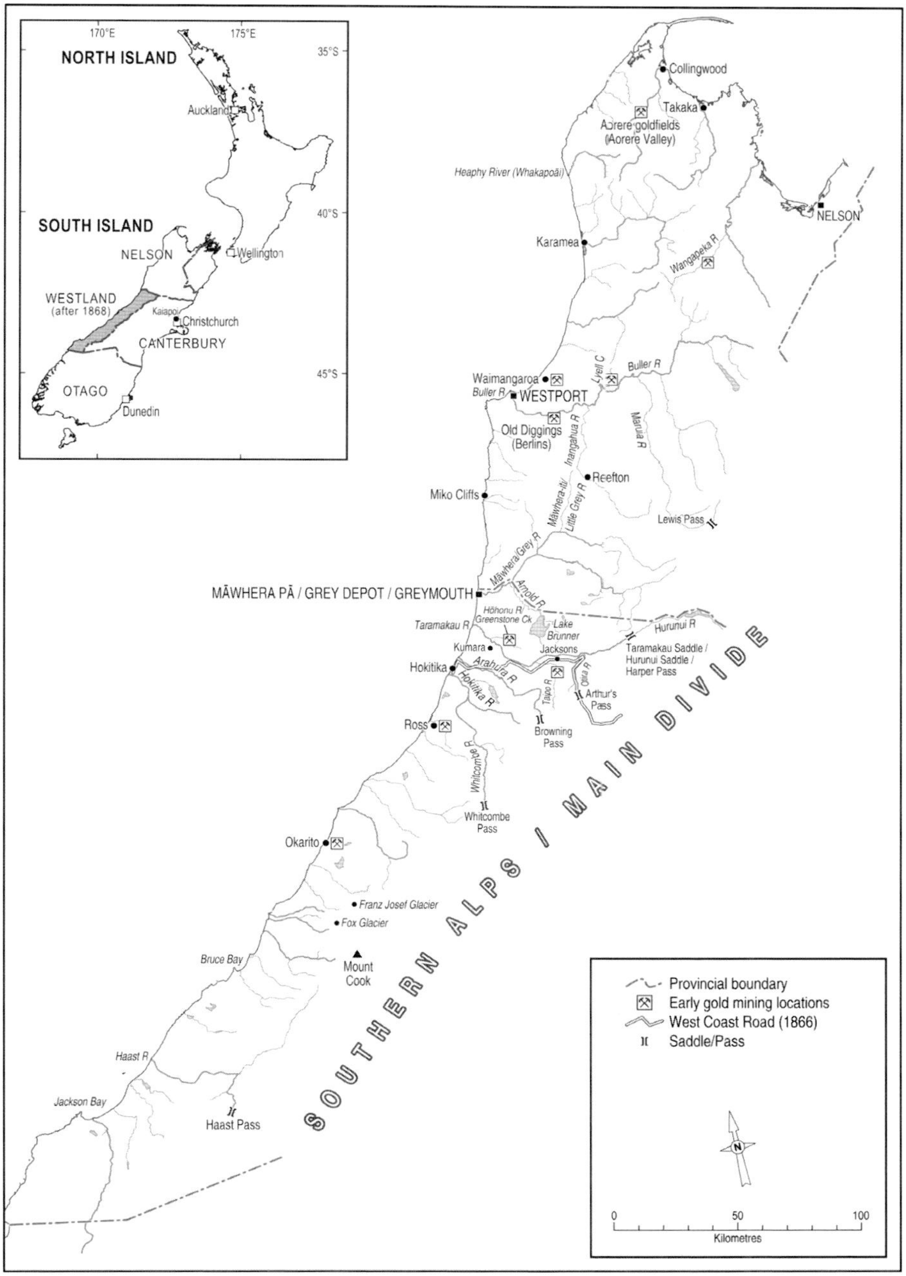


The West Coast and Nelson Regions covered in William Smart's 'The Westland Goldfields'.

Map by Tim Nolan, based on Tony Nolan, *Gold Trails of the West Coast*, Reed, 1975, p. 2.

# PART 1

# In search of that golden land

*Hark: What sound is that I hear*
*Ringing so plainly on the ear?*
*'Tis the march; 'tis the march of the bold pioneer*
*In search of that Golden Land.*[1]

Most of what we know about William Smart we learn from the way he represents himself; we don't see him close-up through the eyes of others. But enough can be gleaned from his writings to reveal a hardy, plain-speaking, enterprising man with an eye to the main chance. His early years in Canterbury followed the pattern of many other young gentlemen in the colony; but his obsession with gold, and the fact that he recorded, in both words and pictures, his search for it in remote and rugged settings, make a remarkable story.

Smart's own introduction to 'The Westland Goldfields' gives some background to those times and places, but it is limited and rather fragmented. This chapter therefore provides some context for Smart's account. The bigger story of the gold rushes in Nelson and Canterbury provinces is well documented elsewhere.[2]

In the 1860s, when these events took place, New Zealand was divided into self-governing provinces. The western side of the South Island was split between the three main southern provinces: Nelson, Canterbury and Otago. The territory north of the Grey River was part of Nelson; Fiordland belonged to Otago; and the middle section, between the Grey River and Big Bay in the south, belonged to Canterbury, and was known as West Canterbury.

By the late 1880s, when Smart was writing his account, West Canterbury had become the 'Westland' of his title, having won its independence from Canterbury in 1868 to become a separate county and then, in 1873, a province in its own right. The 'West Coast', on the other hand, is a geographical name for the longer strip of land stretching 600 kilometres from Kahurangi Point in the far north of the island to Big Bay in the south, and from the Tasman Sea to the Main Divide.

The West Coast lies hidden from the rest of the South Island behind the great wall

Early explorers and prospectors on the Coast often could not linger to muse on the landscape. William Smart noted: 'The scenery along the beach line is very grand … but we were too hungry to properly appreciate it.' When time and conditions allowed, however, he would sit and sketch in his diary the scene before him.

'Main Divide of the Southern Alps from the Okarito Trig Station, Westland'. Making New Zealand Centennial Collection. Ref: PAColl-8550-23, Alexander Turnbull Library, Wellington, New Zealand. http://natlib.govt.nz/records/23183052.

of the Southern Alps, which extend almost the entire length of the island. Its isolation, topography and climate have had a profound influence on its history. In the coastal lowlands, dense rainforest, low scrub and thick tangled undergrowth blanketed the land from the snowline to the edge of the ocean. Sandhills, river flats, lakes, swamps, hills, deep gullies and the occasional expanse of open plain provided some variation to this otherwise unbroken landscape. These lowlands were crossed by many creeks and rivers that drained the valleys. Above towered the spectacular tiers of snow-covered mountains, and two glaciers curved down almost to sea level from the Main Divide. The beaches were littered with huge piles of driftwood: whole trees were swept down from the hinterland in frequent floods. The pounding surf constantly made and remade the sandbars at each river mouth; these were always treacherous for ships, and the coastline became littered with wrecks. This was, and still can be, a wild place, and William Smart saw the drama and beauty in it.

Smart's close bond with the Poutini Ngāi Tahu people of Māwhera, and their early involvement in gold mining, mean that they were an integral part of his story. Māori had been on the Coast – mainly in the north – almost from the time of their arrival in New Zealand. Their settlements on the Coast were located mainly at river mouths or estuaries; the ones that feature in this story were at Kawatiri (Buller River), Māwhera (Grey River)[3] and the Taramakau.[4] Their numbers cannot be accurately determined but were never large, in contrast to the more populous tribes of the North Island and the north and east of the South Island. Māwhera chief Tarapuhi te Kaukihi told James Mackay in 1857 that there were 87 Māori on the West Coast, from Whanganui Inlet in the north to Foveaux Strait, and that he knew most of them by name. He said that 37 of them were in Nelson Province (i.e., north of Māwhera), which left 50 from Māwhera south. An official estimate in 1868 put the numbers between Māwhera and Bruce Bay at 68.[5] After just two decades of colonisation and the establishment of British law, Māori still maintained most of their traditional settlements and tribal ways, but had incorporated some European customs and technologies – including Christianity and, for some, gold mining.

Poutini Ngāi Tahu, a hapū (subtribe) of Ngāi Tahu, the main South Island tribe, held sway over the Coast and were guardians of the region's sources of pounamu (greenstone).[6] Life on the Coast was harsh, and some have suggested that, had it not been for the pounamu, it is doubtful whether Māori would have settled there at all. So highly was it valued that the South Island was known from early times as Te Wahi Pounamu ('the place of greenstone').

Māori were skilled explorers and bush travellers. Their journeys took them to all corners of the South Island, using the beachline, riverbeds and a network of alpine trails. They were known for their long journeys to and from the sources of pounamu in rivers on the Coast and in West Otago (now Fiordland), and to seasonal mahinga kai (food-gathering areas). Through years of experience and oral tradition, they had built up extensive knowledge of the best routes and alpine passes, and when winter snow and ice would disappear.

Tarapuhi told William Smart he knew of numerous trails across the mountains to the east of the island via the Otira and Rakaia rivers and Tioripatea (Haast Pass), but that, for his people, only one was still in regular use – the Taramakau Saddle, between the headwaters of the Taramakau and Hurunui rivers.[7] He said that he and his kin would trek from Māwhera across the pass to Kaiapoi in the east, and that he himself had covered all the country from Nelson to Lake Wanaka.[8]

From the late eighteenth century, sealers and whalers built land stations along the coast, particularly in the south; but by 1840 the slaughter was over, the animal populations were depleted and the settlements abandoned. In the 1840s Thomas Brunner, Charles Heaphy and William Fox were the first Europeans to explore the region. With their Māori guides, they covered the territory from Nelson to Tititira Head, just south of Bruce Bay. It was through their written accounts of these epic expeditions that European settlers first came to know the nature of these western lands.[9]

Elsewhere in New Zealand, colonisation was gathering pace: settlements were

The village of the Tara Makau, Arahura. The settlement stood on the north bank of the Taramakau, 13 kilometres south of Māwhera (Greymouth). Heaphy showed it as it was before the discovery of gold and the influx of European settlers in the 1860s.

'The village of the Tara Makau, Arahura', [wood engraving after Charles Heaphy, 1846] in Sarah Tucker, *The Southern Cross and Southern Crown, or, The Gospel in New Zealand*, 3rd edn, James Nisbet, London, 1856, p. 163.

taking root, roads and ports were being built, provincial governments were formed, and farming, industry and trade were beginning to flourish. The planned settlement of Canterbury was founded in 1850, bringing a steady influx of mainly English colonists to its eastern plains and the new town of Christchurch. William Smart was one of those early arrivals.

William Smart was born in London in 1830, the fourth of six children of Rebecca (née Franklin) and John Smart. His father was a wealthy coal merchant who ran his business from Red Lion Square in Holborn. However, by the time William was six or seven, he and his five siblings had been orphaned and made wards in Chancery. This meant they were under the guardianship of the Court of Chancery and beneficiaries of an inheritance administered by the court, giving each of them financial independence.[10]

Little more is known of Smart's early life, but he clearly received a decent education. On leaving school he was apprenticed to William Cubitt & Company, architects and master builders of London. The company was engaged in large-scale civil engineering and construction projects, including elegant neo-classical residences and well-known public buildings such as the Covent Garden Theatre and market halls, the new façade of Buckingham Palace, and Osborne House on the Isle of Wight for Queen Victoria and Prince Albert. Young Smart appears to have done well: in the 1840s, barely out of his apprenticeship, he became a clerk of works on several of Cubitt's building projects.[11]

In the late 1840s the Canterbury Association was founded in London to establish a settlement in the South Island of New Zealand, sponsored by the Church of England, and began recruiting suitable English men and women to populate it. William and his older brother John Franklin Smart were attracted to the association's vision of a model English colony. For these independent young men it was the chance to own land and to prosper; and it offered the opportunity for a great adventure.

The brothers set sail from Plymouth in October 1850 aboard the *Isabella Hercus*, the sixth of the Canterbury Association's immigrant ships. Their party consisted of William (not yet 21), John (25), his wife Hannah, and her siblings, Richard and Henrietta Hayter. Whereas most of their fellow passengers were in steerage, they travelled 'fore cabin', as young people of some means. The ship berthed in Lyttelton in March 1851.

The Smart brothers quickly took advantage of what Canterbury had to offer. Soon after they arrived they set up in business as timber and firewood suppliers and cartage contractors, and each acquired both town and rural properties. In 1854 William was listed on the Christchurch electoral roll as a 'gentleman' who owned freehold land in Madras Street, from where he conducted his business. Among interests in other rural properties, he leased a sheep run near Kaiapoi, between the Styx and Waimakariri rivers, in partnership with Christchurch butcher Charles Turner; it was known locally as 'Smart and Turner's run'. He also went into partnership with George Day on another run north of the Waimakariri.[12] Like Turner and Day, and friends Richard Hayter and George 'Jockey' Lee of Wharfedale Station, Smart owned and bred racehorses. He was admitted to membership of the elite Canterbury Jockey Club, and became a race-day steward.[13]

In 1860, amid fears that the fighting in Taranaki over disputed Māori land would spread to other parts of the country, all able-bodied men of fighting age were conscripted into provincial militia. William Smart became an active member of No. 5 Company of the Canterbury Rifle Volunteers, based in Kaiapoi, and was unanimously elected to the rank of lieutenant by the troops.[14]

In 1862, when 'The Westland Goldfields' opens, Smart had been in the colony for 11 years. Canterbury's settler community was forging ahead, intent on building an ideal colony that replicated the religious, social and political institutions of England, and a solid foundation for prosperity and progress, with agricultural and pastoral farming at its heart.

West Canterbury, however, was a different story. In the east of the province, the area to the west was little known and largely ignored. General impressions of it were coloured by Thomas Brunner's harrowing accounts of his journeys there: the West Coast he described was a dismal wilderness of mist and rain, deep swamps, impenetrable forests and treacherous rivers – a stark contrast to the relatively benign, open, increasingly populated and cultivated landscape of the Canterbury Plains. In between, the formidable barrier of the Southern Alps accentuated the remoteness and mystery of the territory, which was regarded by many with a kind of 'superstitious dread'.[15] This reputation was difficult to shake off. More than a decade after Brunner's journeys through the region, his estimation of it was still the generally accepted view – that it was of no value to settlers. The territory remained 'far more distant than the North Island and definitely more inaccessible'.[16] Some believed it would be suitable only for a penal colony.[17] During the early 1850s, the only attempt to explore the region was when the ship *Acheron* arrived in 1851 to conduct a marine survey of the coastline. Apart from the occasional vessel that anchored off the Buller River or at Jacksons Bay, no Europeans ventured there.

In the second half of the 1850s, the western region of both Canterbury and Nelson provinces began to attract new interest. The provincial government in Christchurch faced mounting pressure to open up more grazing land, but a proposal to spend £100 to explore the West Coast was rejected as an 'unjustifiable expense'.[18] In 1857, when a new wave of land explorers ventured there, it was as if to unknown territory.

The first approaches were from Nelson. In early 1857, James Mackay led an expedition down the northern section of the Coast as far as the Buller and Grey rivers. He reported that Māwhera had the potential to become a town built on timber milling, coal mining, agriculture and its splendid port. He also identified a quarter of a million acres of open land in the Grey valley, which prompted Smart's friend Jockey Lee to lease part of it for the first sheep run on the Coast, through a deal arranged with Tarapuhi.[19] The Nelson provincial government sent surveyor John Rochfort to the Coast to explore and survey the land. In 1859 the ship *Mary Louisa* sailed from Lyttelton to inspect the West Coast from the sea for possible new grazing land, but the region did not impress. The voyage 'certainly added no item of a favourable character to our information'.[20]

The first Europeans to trek overland from eastern Canterbury to the West Coast were two parties who crossed the Alps via the Taramakau Saddle in 1857, guided by Māori from Kaiapoi. The first party was led by Canterbury's provincial engineer, Edward Dobson; the second by Leonard Harper.

Harper had become curious after reading that some Kaiapoi Māori had fled across the Alps to safety after a war party led by the Ngāti Toa chief Te Rauparaha laid siege to their pā in 1831. In November 1857 Harper persuaded the Kaiapoi chief Wereta Tainui to allow his son, Ihaia Tainui, and two others to guide him and Edwin Lock across the Taramakau Saddle to the West Coast. Unlike Dobson, Harper's party reached Māwhera and explored a long stretch of the coastline, escorted by chief Tarapuhi. Ignoring generations of Māori travellers who had gone before, the Canterbury establishment hailed Harper's coast-to-coast journey as the 'first' crossing of the Alps, and the pass was promptly renamed after him.[21]

Māori had known there was gold in New Zealand long before European settlement. They had noticed it in certain riverbeds, but had no use for it – pounamu was what they prized for its utility and beauty, and for its role in intertribal trade.

Whalers discovered traces of gold in the Coromandel district in the early 1840s, and a survey party found signs in the Takaka River, near Nelson. Both locations later proved to be rich goldfields, but no sustained rush occurred at the time. In the early 1850s, however, the authorities became concerned about the stream of settlers leaving for the Australian goldfields. In 1852 some influential Auckland citizens offered a £100 reward (later increased to £500) for the discovery of gold in payable quantities in that region. Later that same year there was a minor rush to the Coromandel, and the prize was awarded.

The first serious goldmining took place at Aorere, in the part of northwest Nelson Province now called Golden Bay. The first reports of gold there came in October 1856. The Nelson provincial government did not encourage miners to the area,[22] but the word spread; by February 1857 there were about 60 miners in the Aorere and Takaka valleys, and a mining town had sprung up at Collingwood. This and later gold rushes in the area – and some of the people involved – have a strong bearing on Smart's story, as they explain how gold mining first arrived in West Canterbury and who brought it there.

Here, as elsewhere in New Zealand, Māori observed how Pākehā prized this yellow metal. Being generally quick to pick up European ideas and technologies and adapt them to their own purposes, many of them became skilled and enthusiastic gold diggers themselves. Some Māori had worked on the Victorian goldfields during the great rushes of the 1850s[23] and had returned and worked on the early Nelson diggings.[24] One newspaper commented: 'Greatly may it be doubted whether the natives here have so intense a desire for anything as for gold.' Their success was noted – 'Maoris go where Europeans would not go and persevere under difficulties'[25] – and so were their unorthodox methods, such as diving for the gold, and working in groups that included women and children.[26]

The number of diggers at the Aorere soon swelled to over 1300 Europeans and 600 Māori. Many of the Māori had arrived from the North Island. Some had kinship links with local iwi, including 60 who were reported to have come from Porirua, near Wellington.[27] This influx put a strain on the hospitality and resources of chief Tamati Pirimona Marino and his Aorere community.[28] Some settlers were concerned that gold was having a negative effect on Māori life, diverting energies away from food production[29] and other traditional activities, and eclipsing their worship of the Christian God.[30]

By 1859 the main rush was over and most of the diggers, Māori and Pākehā, had moved away. Some returned home; others went south and east from Aorere and Takaka to prospect in Nelson's rugged backcountry. News of progress on the diggings, gold yields and new rushes was reported in local papers and syndicated further afield. But on the diggings, word of mouth was vital in transmitting knowledge of new goldfields, and particularly in passing the information to Māori communities down the West Coast, where no settlers had yet arrived.

One instance of this long-distance communication shows how Poutini Ngāi Tahu at Māwhera, 350 kilometres south of Aorere, heard news of a gold rush. In 1858 Tarapuhi and his brother, Wereta Tainui, travelled inland up the Grey River with Jockey Lee to the sheep run that Lee had just leased from Tarapuhi. On their return east across the Alps, they reported that a group of Māori miners from the Nelson goldfields had told them about rich diggings at a Māori settlement on the north side of the Buller River. An 'immense abundance' of gold was being gathered by both Māori and Pākehā at a place they called 'Whakapoi', reported the *Lyttelton Times*: 'There is of course a great deal of the usual Maori exaggeration in these statements; but the existence of the gold is undoubted, many specimens having reached town.'[31]

The location of this goldfield is not clear: the Whakapoāi (Heaphy River) is 130 kilometres north of the Buller, and no gold rush was reported there. The information was hearsay, and the message appears to have become scrambled. But the incident reveals several things: how the 'bush telegraph' operated between communities up and down the Coast; that Māwhera Māori were well aware of the gold rushes in the north; and that Māori there were actively involved in gold digging, with considerable success.

In 1857 came the first hints of gold in Canterbury's own backyard. When Edward Dobson returned from his journey across the Taramakau Saddle he reported that he had observed quartz that was possibly gold-bearing, near the Main Divide. Harper, who followed Dobson a few months later, announced that he had found traces of gold at the mouth of the Taramakau and Grey rivers, and had brought back to Christchurch some 'fine specimens'. There was no reaction to these signals, though, and any interest in West Canterbury soon waned.[32]

At the end of the 1850s more goldfields were discovered in Nelson's hinterland, but still none had been found on the West Coast. By then, however, the Coast was arousing interest for different reasons. Others now came calling: surveyors and geologists, land purchase agents and pastoral farmers.

Then, in 1859, the random reports and rumours of gold finds were given credibility by two men of science. In August, John Rochfort, surveyor for the Nelson provincial government, reported that on his second expedition to the Buller River he and his survey party had found signs of an abundance of gold: 'the royal mineral was lying on the edge of the river glistening in the sun, and in such quantity as induced a rather mutinous spirit, my hands having a greater preference for the golden prospects before them than the sterner duties of surveying'.[33] His account of the expedition, published in Nelson and beyond, stated that many tributaries of the Buller were without doubt gold-bearing, and he confidently predicted that gold would soon have 'a powerful influence on the future of this long neglected West land'.[34]

The second such opinion came in October from Ferdinand von Hochstetter, a visiting German geologist who had been invited to make surveys of parts of the North Island and in Nelson Province. In a public lecture in Nelson, he suggested that a gold-bearing zone might extend continuously the whole length of 'the Middle Island' (i.e. the South Island); and he identified the Buller River as one site 'for further prospecting where the indications were strong'.[35]

As Rochfort's report spread, prospecting parties were lured to the site. In May 1860 a party of 20 Canterbury prospectors sailed to the Buller on board the *Fantome* to search for it, but the combination of remoteness, inaccessibility and flooding forced them to retreat, disappointed. Another party arrived in the *Gipsy* and were equally unimpressed – one member declared he would not give 'a snap of his fingers' for the whole region – although a few did stay to prospect.[36]

Despite these false starts, it was becoming clear that Canterbury might soon be facing its own gold rush, whether or not it wanted one – and many of the founding fathers did not. But even if there were payable gold within its boundaries, any action would be premature because West Canterbury was still in Māori ownership. Somehow the region had been omitted from the 1848 Kemp Purchase, supposedly of the entire South Island. Until the provincial government could confirm with the central government, then based in Auckland, that Māori title had been extinguished, the declaration of any goldfield in West Canterbury could not be considered.

In 1860 James Mackay completed the purchase of virtually the whole of the West Coast from Poutini Ngāi Tahu on behalf of the Crown – known as the Arahura Purchase – for a miserly £300. It was later revealed that Mackay and other central government officials had known there was gold in the region, which naturally increased the land's value, yet any reference to gold was withheld from negotiations; Mackay said not a word about it until the purchase was completed.[37] In Christchurch the transaction stirred little interest, as did any ideas about the settlement or development of the region.

Meanwhile in Nelson Province, new gold discoveries were moving south. Rushes had already broken out at Aorere and Takaka, and more recently at the Wangapeka River. Now the Buller district would be added to the list. Like many of the other Nelson rushes, those at the Buller were at first an almost exclusively Māori affair. The first find was 40 kilometres up the Buller River, close to present-day Berlins; it later became known as the 'Old Diggings'. As the story goes, some Māori working on the Aorere diggings recalled having previously seen gold on the banks of the Buller, and their new knowledge of gold mining led them back to the site.[38] Another, possibly related, version was that the rush to the Old Diggings may have been started by members of John Rochfort's 1859 survey team. There were 10 Māori from Māwhera in his party,[39] and they now knew that the 'royal mineral' was to be found in great quantity on the Buller, and exactly where it was located; and it was to this spot that they returned later.

Among those working at the Old Diggings in 1861 was Ihaia Tainui,[40] one of the Kaiapoi Māori who had guided Harper across the Taramakau Saddle in 1857.

Afterwards he had gone to live at Māwhera with his uncle, Tarapuhi, who had known about the gold rushes at Aorere since at least 1858. It seems likely that this is where Ihaia learned the techniques of washing gold.[41] It also seems likely that he was one of the Māwhera men in Rochfort's survey team. Ihaia Tainui and other Māori miners had been working undisturbed on these first Buller diggings for some time when the newspapers got word of it in April 1861. A breathless Collingwood reporter broke the news:

> I write to you in haste to tell you there are some Maoris arrived from the Buller river, West Coast, with a quantity of gold … Dr. Mackley[42] has returned and brought some gold with him, and another Maori is expected daily with upwards of a pound weight. This news has given a stir to all our Maoris who … are arranging for an immediate rush to the scene. Should the account be true – and there is every appearance of it being so – we shall have the white men speedily following them.[43]

In May and June, as Māori began arriving at Collingwood from the Buller diggings with hauls of up to 52 ounces of gold,[44] more miners were attracted to the new location, despite its remoteness, the scarcity of food and the perils of the river. Tainui was uneasy about a rush of newcomers, especially Pākehā, and expressed his fears in a letter (*see opposite page*) to James Mackay, Assistant Secretary of Native Affairs, who was stationed at Collingwood.

Reuben Waite now enters the story. He was typical of the seasoned storekeepers who played a pivotal role in every gold rush. When a new goldfield was discovered, one or two storekeepers would travel there to set up shop. The miners followed, knowing there would be food supplies to keep them alive and a buyer for their gold until the bankers arrived. Many storekeepers would 'grub stake' prospecting parties, supplying needy diggers with provisions to enable them to try new ground which, if shown to be promising, would attract more miners, cause a rush and in turn boost the storekeeper's profits.

At this time, Waite was in business at Collingwood; he had left London some years earlier and followed the gold rushes to California, Victoria and on to New Zealand.[45] When a group of Māori diggers arrived at his store with splendid pieces of Buller gold, he saw a new opportunity. He chartered a ship, ordered provisions, tools and equipment, and set sail for the Buller, where he opened the first store near the river mouth and began trading. These small beginnings at the Old Diggings, far up the river, marked the first of the West Coast gold rushes and the establishment of Westport, and earned Waite the status of 'father of storekeeping in the west country'.[46]

In July 1861, news spread about a more accessible goldfield on the coast, 16 kilometres north of the Buller River, at Waimangaroa. Yet another rush broke out,

Te Puna Kawatiri
24 April 1861

… to Mr Mackay,
Friend, salutation to you. This is our letter about the Diggings at Kawatiri (River Buller) …
These are our words which you are to listen or attend to, viz:

1st Do not permit the custom of murdering men to come to the Diggings.

2nd Do not allow disputes about claims to arise at these Diggings. Let us have peace only.

3rd Do not allow stealers of gold, or nuggets, or of rich claims to come to these Diggings.

4th Lest evil should arise at these Diggings.

5th O Maoris who reside on this Island [and on Northern Island] on the one side and on the other of it. Come in peace to this work and bring with you love alone.

…

7th O Europeans who are inhabiting the Towns of this Island [and the towns of the Northern Island], come in peace and do not bring with you to these diggings the custom of killing men, for when Pakehas dug for gold in their Islands, Port Phillip [Victoria] and California they were in the habit of killing men.

…

9th Do not permit guns and powder to be brought to these diggings, lest the Europeans and the Natives work in fear of guns and powder, let them bring but one gun and powder, [and] a knife to scrape the gold out of crevices of the ledges of rocks.

10th If either the Pakehas or Maoris wish to bring guns or powder to these diggings, let the Law send them back, as who knows? or who thinks? (what may happen) as you Europeans have many customs.

This is the whole of our speech to you, if it appears to you to be evil, cast it away, if it is good, work at it (publish it) so that the seed may be well scattered about, that is the words of this letter …

Signed
Na Ihaia Tainui
Na Tapara
Na Poharama Hotu X
Na Hakarama X
Na te Ru X[47]

Letter from Ihaia Tainui and others to James Mackay Jr, 24 April 1861 [translation].

Outward Letters, Assistant Secretary of Native Affairs, 23 February 1858 – 8 July 1863, MA-Co lingwood 2/1, Archives New Zealand, Wellington.

again led by Māori, 'there being but few white men in the district …'[48] Enthusiasm for gold digging was infectious, and more Māori were eager to learn the techniques. In August a group from Māwhera travelled up the coast to the Waimangaroa diggings 'for the purpose of gaining instruction in the mode of digging for and washing gold. When obtained they would then return to the Grey and search for it in that district.'[49] However, the difficult conditions meant that large numbers were never attracted to the early Buller diggings. These rushes merely served as a 'modest prelude' to the great discoveries that were to come further south, in West Canterbury.[50]

Meanwhile, back in May 1861, Central Otago had grabbed the headlines when prospector Gabriel Read made his sensational discovery: he reported that he had seen in the ground before him an astonishing abundance of gold, 'shining like the stars in Orion on a dark, frosty night'.[51] The news triggered a series of major gold rushes across the region. From around New Zealand, Victoria and further afield, a great migration to the rich Otago goldfields began, stealing the limelight from the rushes in Nelson Province.

Canterbury could only look on. As early as 1851, some citizens had seen the effects of the Victorian gold rushes and had begun to fret, especially about the exodus of its population. Now, 10 years later, there was a gold rush on the province's doorstep and, predictably, its menfolk were being drawn away in worrying numbers. Furthermore, Canterbury now had to consider the possibility of a rush within its own borders. For an otherwise sober, hard-working province this was an unwelcome prospect. Such an event would create a 'boom and bust' economy, divert its labour supply, and interfere with the harvest. It would also threaten the province's moral fabric by attracting the usual camp followers: the ruffians, sluts, swindlers, gamblers, reprobates, vagrants and other undesirables.[52] The *Press* warned: 'Like Frankenstein, [the people of Otago] are appalled at the monster they have created.'[53] Canterbury was advised not to follow suit.

For Canterbury's founding fathers, the more prudent road to prosperity lay in taking advantage of other provinces' goldfields by trading with them. This had worked in South Australia, where 'the presence of a neighbouring gold field [in Victoria] was the occasion of enormous prosperity within a very short time'. And as many left for the Otago diggings, those who stayed were told by the *Press* that they would 'be content with the surer reward for ordinary labour'.[54] Indeed, staying in Canterbury became a matter of moral superiority and provincial loyalty. One Lyttelton baker took out a newspaper advertisement to deny rumours that he was off to Otago; he threatened to punish the author of such lies, and to take legal action if the allegations continued.[55]

William Smart had also been following this steady path. By now he had returned to his former occupation as a building contractor, and in his spare time he continued to breed racehorses, develop his sheep run near Kaiapoi and train with the local Rifle Volunteers.[56] But he was not immune to the call of gold. Sometime in the second half of 1861, at the age of 31, he downed tools and joined the flow of men south to Central Otago. He remained there for some months and learned the techniques of gold panning and sluicing, something of goldfields geology, and the art of survival on the diggings.[57]

In the early 1860s, Canterbury's menfolk were deserting steady jobs for an uncertain future on the Otago goldfields. The *Lyttelton Times* asked 'the working men of this province to consider ... whether it is worth their while to incur ... dreadful risks, and fool away their time and money in running after what is a hundred chances to one they will never get'. *Lyttelton Times*, 13 September 1862, 4.

William Hodgkins, 'Off to the diggings, Otago, c1865'. Ref: E-016-2-038, Alexander Turnbull Library, Wellington, New Zealand.

Meanwhile, there was mounting pressure for Canterbury to find its own goldfield. In September 1861 the *Lyttelton Times* conceded that 'a gold field, however dangerous it may be to the peace and happiness of those to whom it is granted, is yet undoubtedly desired by a very large majority'. Furthermore, having its own goldfield would halt the outflow of its citizens to Otago: it was 'impossible to conceive a more powerful method of attracting [and keeping] population'.[58] More incentives were needed.

Back in 1853, a reward of £200 had been raised by public subscription for anyone who found a 'remunerative gold field available to the Canterbury settlement';[59] but it had come to nothing. When the gold rushes began in Nelson Province in 1857, the Canterbury provincial government had increased the reward to £500, in part to prevent the departure of its men for the north. The *Lyttelton Times* was scathing; it described the move as 'reckless': 'If gold exists, it will be found without a reward; if not, it will not be found by the offer.'[60] Ignoring this logic, Cantabrians now called for a new, richer reward. Nelson, Southland, Wellington and Auckland provinces each offered one; and hadn't Otago's reward produced the desired result? Gabriel Read had been tipped to claim it, and later that year the prize was indeed his.[61]

At first the Canterbury government voted down the proposal, but in August 1861 public opinion forced its hand. In a well-planned strategy, promoters of a reward called a public meeting in Christchurch. Canterbury, the crowd was told, could no longer stand quietly by and watch its neighbours boast of their goldfields while it neglected to search for its own. Provincial geologist Julius Haast told the meeting there was little doubt that gold would be found in the headwaters of North Canterbury rivers such as the Hurunui and, particularly, the Waitaki River to the south. He supported the resolution for a new gold reward, and hoped 'to be the first to find the gold himself'. The meeting voted unanimously to petition the government to offer a new reward – for the benefits it would bring to the province, and to arrest the present 'eflux' of population to goldfields elsewhere.[62]

What was a prudent government to do? Having received the petitioners, it reviewed the budget estimates and placed the following advertisement:

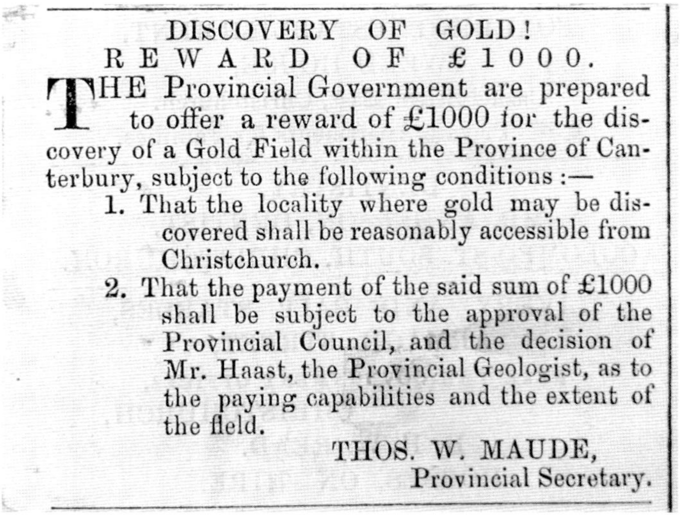

DISCOVERY OF GOLD!
REWARD OF £1000.

THE Provincial Government are prepared to offer a reward of £1000 for the discovery of a Gold Field within the Province of Canterbury, subject to the following conditions:—

1. That the locality where gold may be discovered shall be reasonably accessible from Christchurch.
2. That the payment of the said sum of £1000 shall be subject to the approval of the Provincial Council, and the decision of Mr. Haast, the Provincial Geologist, as to the paying capabilities and the extent of the field.

THOS. W. MAUDE,
Provincial Secretary.

This notice appeared regularly in all three Christchurch papers between September 1861 and March 1863. At £1000 the reward was twice the amount the Canterbury government had previously offered. Wellington province offered the same amount but with tighter conditions, and its definition of 'payability' was clearly spelt out. Nelson's £2000 reward was on a sliding scale, depending on the goldfield's yield within a specified timeframe. *Colonist*, 18 December 1863, 4; 'Available gold field', *Wellington Independent*, 17 May 1864, 3.

Those in Canterbury who longed for their own goldfield looked on enviously as the rush to Otago gathered momentum.

Charles Edward Hammond, 'Pedlars at the diggings, Lindis Gold Field, Otago'. Making New Zealand Centennial Collection. Ref: MNZ-0425-1/4-F. Alexander Turnbull Library, Wellington, New Zealand. http://natlib.govt.nz/records/23130C75.

Both the government and the press assumed that Canterbury's goldfield would be found in the east of the province, close to Christchurch, where it would be convenient to administer and of direct benefit to the local economy. Its discovery was seen as 'probable' by the close of the year.[63] This was wishful thinking, of course, aided and abetted by Haast. Almost daily, fresh reports appeared in Christchurch newspapers of the riches washed from the Otago goldfields. In early 1861 gold had been found near Lindis Pass, close to the border between Otago and Canterbury. This find appeared promising, but had been overshadowed by the more lucrative finds at Gabriel's Gully.

During that year Haast, in his role as provincial geologist, had explored the Rangitata and Ashburton rivers and in the Malvern Hills, about 65 kilometres west of Christchurch, always on the lookout for gold. The people of Canterbury put great faith in his fieldwork, but he was unable to produce any evidence of the 'glittering talisman'. He did, however, find 'something else not a whit less valuable' in the Malvern Hills: an extensive seam of high-quality coal.[64] This was good news, of course; but it was not what the province lusted after.

In the first quarter of 1862, Haast was conducting a geological survey of the Mackenzie Country and was again prospecting. He did find likely-looking schist rock, similar to that found at the Otago diggings, but despite his earlier confidence he had to admit he had again failed to find any gold.[65]

William Smart had meanwhile been trying his luck on the Central Otago diggings; but by May 1862 he was back home in Kaiapoi, setting up in business with George Masterton Clark. The partnership advertised a wide range of useful services: 'Messrs Smart and Clark, architects, accountants, house, land, stock and general commission agents, auctioneers and valuers'.[66] Apparently he was returning to a settled life.

By the middle of 1862 nobody had yet claimed Canterbury's gold reward, although many felt it was just a matter of time. Some even blamed Haast for not finding *their* goldfield. Then in August came the first application. It was from a farmer near Christchurch, who found what looked like traces of gold in the gizzard of a duck. All the ducks in the flock were slaughtered, and their gizzards were found to contain golden specks. But after the initial excitement and some digging in the area, nothing else was found and the claim lapsed.[67]

In August a second gold rush broke out near the Lindis Pass, at a goldfield known as 'Hartley's'.[68] Surely gold would be found nearby, on the Canterbury side of the provincial boundary.

Fresh measures were needed, and once again public pressure was brought to bear. Another deputation of citizens urged the Canterbury government to take the initiative and organise some prospecting parties.[69] Finally, the government relented: it allocated £250 in the budget and placed another advertisement.[70]

**PROSPECTING FOR GOLD.**

THE Provincial Government are desirous of fitting out two prospecting parties, one to go to the north and the other to the south, the latter will be placed under the direction of Mr. Haast. Further information and particulars will be ready on and after Thursday next, the 4th instant. Application to be made to the office of the Provincial Secretray, Christchurch.

THOS. WM. MAUDE,
3340 Provincial Secretary.

*Lyttelton Times*, 3 September 1862, 5.

Charlton Howitt was among those who responded to the advertisement.[71] He accepted the contract to lead the northern expedition, which included a bonus of £250 'in the event of gold being found in payable quantities'.[72] Ten days later, with a party of five, he set out from Christchurch to prospect in the Lake Sumner and Hurunui River region in North Canterbury, and to begin the task of cutting a track across the Taramakau Saddle to the west.[73]

The other party, led by Haast, headed south to inspect the Waitaki district near the boundary with Otago. However, the provincial government just then received another claim for the reward for a gold find in the Opawa Stream near Timaru, and Haast was persuaded to make a detour to investigate. He took the discovery very seriously and made geological pronouncements about the origins of the gold. But to his great annoyance, he realised the gold had been planted in the stream and that he was the object of a 'discreditable' (but to others 'amusing') hoax. He returned immediately to Christchurch in disgust.[74]

It was now October and there was still no goldfield. John Thomas Morris of Timaru summed up Canterbury's continuing frustration perfectly:

THE LAMENT OF CANTERBURY

DEAR Mr. Editor, I pray give ear
To my sad tale, my pitiful lament;
I've not been so annoy'd this many a year,
My inside's really quite in a ferment.
Those other provinces – oh dear, – oh dear,
Are filling me with shame and discontent;
I who was once so vigorous and bold,
Am in the dumps – and why? – *I've got no Gold!*

There's that Otago – oatmeal eating place,
Is turning gold up by the hundred weight.
That fag-end of the world, – now in the race,
Is going ahead at a confounded rate.
See what a grin lights up each Scotchman's face,
When they compare their own with my sad fate.
Otago may be bleak, and wet, and cold,
But what of that when they've got lots of gold.
...
I've got a great geologist to look
About for gold and try can he discern it,
And he's been fossicking in every nook,
But nothing found, or I should quickly learn it –
Except some coal as black as any rook,
And famous fuel – if one could only burn it.
Coal is a mine of wealth, as I've been told
But blow black diamonds – what I want is gold!
...

J.T.M.[75]

It was at this point that William Smart joined the hunt for Canterbury's gold. We do not know how successful his endeavours on the Otago diggings had been, but he had obviously become hooked. There were warnings in the press about the dangers of gold fever, which drove men to

> throw up steady and remunerative employment at a moment's notice to search for a problematical reward in the face of certain and severe hardships … Those who have suffered from the epidemic appear unable to gather wisdom from experience, and are liable at any moment to relapse.[76]

If Smart read this, he took no heed. He had no wife and children, and simply walked away from his life in Kaiapoi. The new business partnership had lasted barely six months. He appears to have left his share of the company's assets to George Clark – along with the debts, which it seems were the greater.

Smart left town and headed into the Canterbury ranges with a party of five. From October to December 1862 they scoured the inland valleys and riverbeds between the Waitaki in the south and the Hurunui in the north, where Dobson had reported likely-looking quartz and where Howitt was now prospecting.[77] Other parties of independent prospectors also went out into the hills in search of Canterbury's gold.

Rumours of gold in the east had come and gone, and expectations that Canterbury's goldfield would be conveniently located near Christchurch were fading: people were reluctantly realising that the focus might have to shift to the far side of the Alps.

In December, in advance of a planned survey expedition to West Canterbury, the government extended Howitt's contract. His instructions were to continue cutting the bridle track across the Saddle, down the Taramakau and across to Lake Brunner; then to make two routes to the sea, one leading down the Hōhonu and Taramakau rivers, and a second down the Arnold to the mouth of the Grey River.[78]

Then, suddenly, the first serious applications for the gold reward came simultaneously from two separate claimants, both from West Canterbury. The story began at the start of October 1862 when the schooner *Emerald Isle* sailed from Christchurch for the West Coast. At the helm was Captain Thomas Dixon, who had been on the California goldfields and was interested in looking for gold. Also on board were Mrs Dixon; 'two young gentlemen' – Alexander MacKillop and William McFarlane – on the lookout for new grazing land; and a load of cargo for Isaac Freeth, who had taken over Jockey Lee's sheep run in the Grey valley. After calling at Dunedin, Wellington, Picton and Nelson, the schooner finally sailed for the Grey River at the end of November.[79]

However, when he reached the West Coast, Captain Dixon mistakenly took the *Emerald Isle* across the bar of the Taramakau instead, unaware that the river was in flood. When the water receded the schooner was left high and dry on the riverbank near the lagoon, stuck fast until the next flood. While they waited, Dixon and some of his crew tried a little gold prospecting – with some success. 'I noticed that the Maoris seemed very much interested in what I was doing,' Dixon commented later. 'I thought they knew nothing of gold washing and it was only idle curiosity.'[80] But one of the observers was Ihaia Tainui, who had returned to Māwhera from the Buller diggings some months earlier to prospect at the Taramakau. He and a mate helped Dixon to refloat the *Emerald Isle*.[81]

Dixon, now impatient to get word of his discovery to the provincial government, and unable to spare any of his own crew, 'asked about for a man' to carry his news and an application for the gold reward overland to Christchurch, together with a sample of his gold as evidence. According to early gold prospector William Martin, who wrote of these events, Tainui 'divined his purpose and offered to go'.[82] The package was to be delivered to Dixon's brother-in-law, John Oakes, proprietor of the Golden Fleece Hotel on Colombo Street and owner of the *Emerald Isle*.

Martin now takes up the story which, he pointed out, was given to him by others, not by Dixon. His lively account deserves to be quoted in full:

> [Tainui] made the trip over on foot in the short time of six days – a distance of 200 miles … and duly delivered his letter and parcel. The three pounds [Tainui's fee] was promptly paid, and he was entertained that night at the Golden Fleece right royally by Captain John Peter Oakes, who was in a state of excitement at the find and future prospects.
>
> Next morning at ten minutes past ten, Captain Oakes, dressed in his best, entered the provincial buildings with his letter [dated 'Taramakau, West Coast, 7 December 1862'] and gold – the contents of the parcel – and asked to see the Provincial Secretary on important business, and when told he was engaged, he asked for Mr. Moorhouse, the Superintendent. He was told that the Super was engaged too. 'The fact is,' said the Official, 'gold has at last been found, and the officials are closeted with the lucky discoverer.' 'That be d___d,' roared Oakes, 'I am the discoverer of the gold, or at least my brother-in-law Captain Dixon is, and here is his gold, and here is his written claim. There is some fraud here, and I will see what it is.' With that he rushed past the man and entered the Superintendent's office.
>
> There he found the Chief Provincial Officers in a state of excitement, looking at some splendid samples of shotty gold[83] and extracting information from the finder who had verbally claimed the two thousand [sic] pounds reward. Oakes was not a little taken aback when he found Dixon's messenger 'Tainui' had forestalled him by ten minutes and was also a claimant for what he deemed to be without doubt his and Dixon's. He declared Tainui had stolen the gold from Dixon's parcel, but on comparing the two samples it was found that while Dixon's sample was fine – some of it as fine as flour – and mixed with black sand, the Maori's was coarse gold, from a quarter to one pennyweight pieces, and no black sand with it; and

> that the parcel Oakes got from his [brother-in-law] was sealed and quite up to the weight mentioned by him.
>
> Captain Oakes raved and stamped about, declaring the Maori a fraud – that no one but Dixon was entitled to the money. The Superintendent pointed out the conditions to be fulfilled before anyone could get the reward, and that was, that so much gold would have to be produced in a given time, and the best thing he could do would be to develop his find … With that he had to be content, and retired.[84]

Who had arrived first at the government offices was hotly disputed. Oakes claimed that he did;[85] but the fact remained that Tainui's claim was registered before Oakes', thus corroborating Martin's account. In the end, neither party was satisfied. Both applications were rejected for not fulfilling the conditions of the reward: the location was not 'reasonably accessible to Christchurch', and neither claimant could prove that their goldfield would pay.

Despite this setback, Dixon boasted that the site would become the richest diggings in New Zealand. Richard Sherrin,[86] however, was sceptical: 'Dixon, having succeeded in a few days … in gathering some 30 or 40 grains of gold, concluded that he had found a gold field, and had nothing to do but make the fact known to ensure the bonus of £1000 offered by the Provincial Government …' But Dixon was not alone in making this assumption.[87]

The incident was widely reported and caused a brief sensation in Christchurch. In January news reached the town that Charlton Howitt had visited the lower reaches of the Taramakau in early December, and declared that the area 'had every appearance of being a goldfield'.[88]

While no reward had been granted, the publicity served to draw attention to the Taramakau as the possible location of Canterbury's goldfield. But to many in Christchurch, the discovery was irrelevant, 'unconnected for any practical purposes with any future gold fields that may be discovered east of the snowy ranges'.[89]

William Smart, however, had become convinced that West Canterbury was where the gold was. He and his party set out for the Coast in December, through the North Canterbury hills, via Mr Taylor's sheep run, past Lake Sumner and across the Taramakau Saddle; for part of the journey they followed Charlton Howitt's new track. About 80 kilometres west of the Saddle they reached the Hōhonu, which at that time flowed into the Taramakau from the north side, about 10 kilometres from the sea. The Hōhonu River has had several names over the years. Its original Māori name was Hokonui, but Pākehā knew it as the Hōhonu or Little Hōhonu (with various spellings) – referring to its source in the nearby Hōhonu Range (see map on page 92).[90]

In early January 1863 there were few people, Māori or Pākehā, in West Canterbury. The main human activity on the Coast was centred on the new settlement of Westport in Nelson Province, the hub of several goldfields along the coast and up the Buller River, where a new rush had just broken out. In November 1862 Māori prospectors had found gold in large quantities 40 kilometres up the river at Lyell Creek, beyond the Old Diggings, and predicted a big rush there.[91] After Tainui's discovery of gold in the Taramakau area, groups of Māori (not all local) were also at work there.

As Smart and his party made their way west, others too were on the trail. Those ahead of him were going to the Buller diggings, but the first few were now beginning to head for the Taramakau. Among them were the adventurers Charles Money and Rowland Davis. Money observed later that the Dixon/Tainui affair 'had excited the minds of the energetic Canterbury folks to such an extent that parties had already left Christchurch and Kaiapoe [sic] to follow in our tracks …'[92] But the numbers were small enough that encountering other travellers was still noteworthy.

Early arrivals at the Taramakau were disappointed, however: food was unobtainable, the conditions were harsh, and barely any gold was found. One report told of 'two men returning back across the saddle almost delirious for want of food'; others arrived at the Buller 'in a deplorable state' after testing Dixon's purported goldfield, complaining bitterly that they had been on a wild goose chase. The men remaining at the Taramakau told Sherrin they considered the goldfield to be little more than a hoax, and some even threatened vengeance: 'It is a shameful thing to lead men astray in this manner, there's no telling what the consequences may be.'[93]

Government men now arrived on the scene. Smart knew that Charlton Howitt's work party had reached the Taramakau Saddle just before Christmas, and in the New Year would continue cutting a track to the sea: they would set up their camp at Lake Brunner. In January 1863 the government's first land survey of West Canterbury began. The party set out from Christchurch and descended the Taramakau River soon after Smart. The survey was led by James Drake, assisted by brothers Richard and William Sherrin, and Swiss adventurer Jakob Lauper (who had only recently been saved from starvation and near collapse on the banks of the Taramakau by 'good samaritans' Money and Davis). In addition, like most official survey parties at the time, Drake's included a Māori guide – Horomona ('Solomon') Tuapaki. The party spent two months surveying the territory between the Grey and Hokitika rivers, and Smart recorded meeting them.

While most of the first Pākehā prospectors were soon forced to leave the Taramakau, others were still arriving. Many, being 'new hands in the bush', were woefully ill prepared. Charlton Howitt noted several parties passing by and worried how they would fare: 'People seem to travel about in this West Coast as though there were a store every few miles.'[94] Richard Sherrin met a party of four staggering up the Taramakau, 'hungry, knocked-up and destitute of food'. Reuben Waite recalled that some 'poor fellows were in such an emaciated state that they were scarcely able to crawl. It was pitiable to see them.'[95] Lauper's earlier plight was reported in Christchurch, and held up as a public lesson: '[The West Coast] is no place for people who have no experience, to try the diggings with the hope of acquiring a speedy pile … should they go there they will stand a good chance … of losing their lives.'[96] Some, like Lauper, were rescued; others were not so fortunate.

When Smart and his party arrived on the Coast they still depended for their survival on the last of their provisions, what nature provided – and the generosity of Māori at Māwhera pā. The settlement stood near the beach on the south bank of the Grey River, about 14 kilometres north of the Taramakau. The number of people living there at any one time varied, but it was never more than a few dozen. It was a mobile population: they covered long distances, travelling by waka on lakes and rivers (the ocean was too dangerous), or on foot along the Coast, inland and across the Alps. Well-proven knowledge and skills made their journeys possible: they left caches of food on the trail; and they knew where and how to find food, and how to make a mōkihi (raft), for example.

Smart was a welcome visitor at the pā and gave a good description of it. Chief Tarapuhi was an old man, probably nearly 70, when Smart arrived there, but he was still deeply involved in tribal affairs and was an energetic long-distance traveller. His hospitality and kindness were legendary; he befriended virtually every Pākehā who entered the pā, and was keen to engage them in conversation. Although food was sometimes in short supply, the obligations of hospitality, observed to a fault by Tarapuhi, meant he shared what he had with those who passed through. Another early traveller described the sort of reception Smart's party would have received. After the welcome cry of 'Haere mai!, Haere mai!',

> dried whitebait, smoked eels and potatoes … were freely offered and accepted by the famished white men … [Tarapuhi's] blanket, half his provisions, the use of his canoes, his own invaluable services were ever placed at the disposal of the Pakeha, free of charge.[97]

In the early months of 1863 Captain Dixon was busy plastering Christchurch with posters advertising the *Emerald Isle*'s return voyage to 'his' goldfield on the Taramakau, and inviting passengers to join him in opening up 'a second California' on the West Coast.[98] The Canterbury government was making arrangements to establish a depot and general store on the Coast. The experiences of early travellers and the expected increase in government activity there had persuaded it to take responsibility for feeding and supporting its own men in the field; to assist needy travellers, after so many accounts of destitution; and to reduce demands on the generosity of Tarapuhi and his people.

Smart's arrival on the Coast coincided with Drake's survey expedition – and two months of unusually settled weather. The glorious summer seduced Richard Sherrin into writing, in the Christchurch *Press*, a glowing account of the region's climate, landscape and future prospects. The editor commented that in Sherrin's report 'we have a picture of the terrible West Coast entirely different from what has hitherto presented itself to our minds. An entirely new country opens before us …'[99]

But the sunshine could not last. The Coast's true identity soon reasserted itself, and Sherrin had to recant. Many of the 'new chums' to arrive on the Coast soon gave up and returned to civilisation, but the hardiest stayed and adapted. They got used to the rain and flooding, to living rough and labouring hard. They came to know the lie of the land and how to traverse it, how to deal with dangers, and how to survive. Much of this they learned from the local Māori, with whom they developed a cordial and co-operative relationship, picking up some of their language along the way.

Smart was one who stayed and adapted. On the Coast he never complained about the conditions, except perhaps the bush lawyer and the sandflies. He relished the life, the freedom, and the pursuit of the 'royal mineral'. His grievances would come later.

The story is now his to tell.

## Notes

### Abbreviations

ANZ Archives New Zealand
ATL Alexander Turnbull Library, National Library of New Zealand, Wellington
ICPS Inward Correspondence of the Provincial Secretary (Canterbury)

1 Attributed to Dan Moloney of Westport (1875–1949).
2 For example: J. H. M. Salmon, *The History of Goldmining in New Zealand*, Government Printer, Wellington, 1963; Cadsonbury, Christchurch, 1996; Philip Ross May, *The West Coast Gold Rushes*, Pegasus, Christchurch, 1962.
3 Māwhera is the Māori name for both this river system and the pā (settlement) at its mouth. Explorer Thomas Brunner renamed the river after Sir George Grey, Governor of New Zealand. In Ngāi Tahu's Treaty settlement of 1997, the river was officially renamed Grey River/Māwheranui. For clarity, 'Māwhera' is used in this text to refer only to the Māori settlement.
4 James Mackay Jr, cited in Philip Ross May, 'The West Canterbury gold rush to 1865: The analysis of a gold rush in the middle of the nineteenth century', unpublished thesis, Canterbury University College, 1953, p. [19a].
5 James Mackay Jr, 'Notes on a journey to Mawera, or Grey River on the West Coast, February and March 1857', *Nelson Examiner & New Zealand Chronicle*, 26 August 1857, 3; May, 'West Canterbury gold rush to 1865, p. [19a].
6 May, 'West Canterbury gold rush to 1865', p. 20.
7 In Smart's time it was generally known as the Taramakau (with variant spellings) or, after the start of the gold rushes, the Hurunui Saddle. From 1857 it was officially 'Harper Pass', but few people appeared to use this name.
8 William Smart, 'The Westland Goldfields, from the diary kept by Wm Smart, one of the discoverers of payable gold in that part of N. Zealand', 1890, Microfilm 64/310, Macmillan Brown Library, University of Canterbury, Christchurch, p. 104.
9 Charles Heaphy, 'Account of an exploring expedition to the S.W. of Nelson', *Nelson Examiner & New Zealand Chronicle*, 14 March 1846, 3, 5; Charles Heaphy, 'Notes of an expedition to Kawatiri and Araura [Arahura] on the western coast of the Middle Island, performed by Messrs Heaphy and Brunner', *Nelson Examiner & New Zealand Chronicle*, 5 September 1846, 105, to 17 October 1846, 130 (weekly instalments); Thomas Brunner, *Journal of an Expedition to the Interior of the Middle Island of New Zealand,* William Clowes & Sons, London, 1850.
10 International Genealogical Index (IGI), https://www.familysearch.org; 'Last will and testament of John Smart of St Pancras, London', 1833, 11/1826, National Archives, London; 'Obituary: William Smart', *Star*, 12 October 1900, 3.
11 William Smart to Superintendent of Canterbury, 8 August 1871, ICPS 1044/1971, ANZ, Christchurch; 'Obituary: William Smart', *Star*, 12 October 1900, 3.
12 L. G. D. Acland, *Early Canterbury Runs: Containing the first, second and third (new) series*, Whitcombe & Tombs, Christchurch, 1946, p. 48. In the 1860s Smart sold his share to Turner; D. N. Hawkins, *Beyond the Waimakariri: A regional history*, Whitcombe & Tombs/Cadsonbury, Christchurch, 2002; 'Electoral Roll for the town of Christchurch', *Lyttelton Times*, 12 August 1854, 4; 'Electoral Roll for the District of Christchurch', *Lyttelton Times*, 16 August 1854, 3; *Canterbury Land Index, 1853–54*, http://www.nzpictures.co.nz/canterburylandindex1853-1864.pdf.
13 *Lyttelton Times*, 23 December 1854, 1; *Lyttelton Times*, 22 December 1860, 5.
14 *Lyttelton Times*, 12 September 1860, 4.
15 *Press*, 24 March 1863, 1.
16 May, *West Coast Gold Rushes*, pp. 61–62.
17 *Lyttelton Times*, 7 April 1864, 4.
18 Ross (Provincial Clerk) to the Superintendent, 'Resolution of the Provincial Council: £100 towards exploration of West Coast', 7 July 1855, CAAR 19936 CH287 ICPS 847/1855, ANZ, Christchurch.
19 James Mackay Jr, 'Notes on a journey from Pakawau, Massacre Bay, to Mawera', *Nelson Examiner & New Zealand Chronicle*, 26 August 1857, 3; *Lyttelton Times*, 3 April 1858, 4; May, 'West Canterbury gold rush to 1865', p. 29.

20 *Lyttelton Times*, 6 July 1859, 4.
21 John Pascoe, 'The Maori and the mountains: Part 2: Across the Main Divide', *Te Ao Hou*, 12 (1955), 26–27, http://teaohou.natlib.govt.nz/journals/teaohou/issue/Mao12TeA/c19.html, accessed 30 January 2016.
22 'Early traces', in A. H. McLintock (ed.), *An Encyclopaedia of New Zealand*, 1966: *Te Ara – the Encyclopedia of New Zealand*, updated 22 April 2009, www.teara.govt.nz/en/1966/gold-discoveries, accessed 25 January 2016.
23 James Cowan, *The New Zealand Wars: A history of the Maori campaign and the pioneering period, Vol. 1, 1845-1864*, Government Printer, Wellington, 1955, pp. 374, 376; Selwyn Korzelinski, quoted in Hilary and John Mitchell, *Te Tau Ihu o te Waka: A history of Māori of Nelson and Marlborough, Vol. 2: Te Ara Hou: The new society*, Huia Publishers, Wellington / Wakatū Incorporation, Nelson, 2007, p. 299.
24 Julia Bradshaw (ed.), *The Diggers' Story: Accounts of the West Coast gold rushes* (An annotated and illustrated new edition of the classic publication, compiled by Carl Pfaff and published by the West Coasters' Association in 1914), Canterbury University Press, Christchurch, 2014, p. 41.
25 *Colonist*, 29 December 1857, 2; Margaret Brown, 'Christmas 1862 at the mouth of the Buller River', *Nelson Historical Society Journal*, vol. 3, 4 (September 1978), http://nzetc.victoria.ac.nz/tm/scholarly/tei-NHSJ03_04-t1-body1-d2-d3.html, accessed 20 January 2016.
26 Patrick Galvin, *Handbook of New Zealand Mines*, New Zealand Mines Department, Government Printer, Wellington, 1887, pp. 238–46.
27 J. N. W. Newport, *Collingwood: A history of the area from earliest days to 1912*, Caxton, Christchurch, 1971, p. 32.
28 D. S. Loder, 'History of the gold discoveries and rushes of the west coast of Nelson and an estimate of their significance', unpublished thesis, Otago University, 1935, Micro-0276a, ATL, p. 15; 'Māori and gold', http://www.theprow.org.nz/maori/maori-and-gold/#, accessed 20 January 2016; *Lyttelton Times*, 29 August 1857, 4.
29 James Mackay Jr, Letter, 22 March 1861, Outwards Letters, Assistant Secretary of Native Affairs, MA-Collingwood 2/1, ANZ, Wellington.
30 Mitchell, *Te Tau Ihu*, *Vol. 2*, p. 307.
31 *Lyttelton Times*, 3 April 1858, 4.
32 *Lyttelton Times*, 26 September 1857, 5; 'Across the Southern Alps', *Press*, 3 July 1909, 6; May, 'West Canterbury gold rush to 1865', pp. 27–28.
33 John Rochfort, 'A brief account of an expedition to the West Coast, December, 1859', *Nelson Examiner & New Zealand Chronicle*, 24 December 1859, 3.
34 Rochfort, 'A brief account of an expedition ...', 3.
35 Ferdinand Hochstetter, 'Lecture on the geology of the province of Nelson', *Nelson Examiner & New Zealand Chronicle*, 1 October 1859, 1.
36 *Nelson Examiner & New Zealand Chronicle*, 30 June 1860, 3; *Lyttelton Times*, 21 July 1860, 5.
37 The Waitangi Tribunal found in 1998 that 'By offering no more than a nominal price for the land which had the potential for a very early substantial rise in value ... the Crown failed to act with the degree of good faith' – which amounted to the Crown being in breach of its Treaty obligation. *Nelson Examiner & New Zealand Chronicle*, 26 August 1857, 3; *Lyttelton Times*, 1 October 1857, 3; May, *The West Coast Gold Rushes*, p. 70n; Waitangi Tribunal, *Ngāi Tahu Report 1991, WAI 27: 13 Arahura*, http://www.justice.govt.nz/tribunals/waitangi-tribunal/Reports/wai0027%201991%20Report/doc_014, accessed 4 January 2016.
38 Newport, *Collingwood*, p. 32.
39 Rochfort, 'A brief account of an expedition ...', 3.
40 Ihaia Tainui (also known as Isaiah, John or Jack Tainui) was a member of Ngāi Tahu nobility, as the son of Kaiapoi chief Wereta Tainui. He was literate, and a politically astute and active campaigner for peace and justice for his people. He was a lay preacher, was appointed a Native Assessor in 1861, and was elected MP for Southern Maori from 1879 to 1881. His correspondence from the Old Diggings in April 1861 shows that he was a seasoned gold miner. Ihaia Tainui and others to James Mackay Jr, 24 April 1861, Outwards Letters, Assistant Secretary of Native Affairs, 23 February 1858–8 July 1863, MA-Collingwood, 2/1, ANZ, Wellington.
41 *Lyttelton Times*, 3 April 1858, 4; Bradshaw (ed.), *The Diggers' Story*, p. 20n.
42 Samuel Mackley was one of the earliest runholders in the Grey valley. He and his large family made much of their income from trading with the goldfields – as well as sheep farmer, he was a

'hotelkeeper, brewer, butcher and storekeeper'. Partially trained in medicine, he was often referred to as 'Dr Mackley' and was called on to treat the ill and injured. For several years he was the Grey Valley representative on the Nelson Provincial Council.

43 *Colonist*, 30 April 1861, 2.

44 52 ounces = 1.5 kilograms; 'The Buller Gold Field', *Nelson Examiner & New Zealand Chronicle*, 25 May 1861, 2; *Colonist*, 12 July 1861, 1.

45 May, *The West Coast Gold Rushes*, p. 255.

46 Reuben Waite, *Narrative of the Discovery of the West Coast Gold-Fields*, W. H. L. Leech (ed.), J. Hounsell, Nelson, 1869, pp. 3–6; R. A. Loughnan, *The First Gold Discoveries in New Zealand*, John Mackay, Government Printer, Wellington, 1906, p. 59.

47 Plus four other signatories. 'X' denotes 'the mark of'.

48 *Colonist*, 26 July 1861, 2; *Nelson Examiner & New Zealand Chronicle*, 24 August 1861, 2.

49 *Nelson Examiner & New Zealand Chronicle*, 24 August 1861, 2.

50 May, 'West Canterbury gold rush to 1865', p. 39.

51 *Otago Daily Times*, 10 May 1911, 10.

52 May, *The West Coast Gold Rushes*, p. 68; *Colonist*, 29 December 1857, 2.

53 *Press*, 28 September 1861, 2.

54 'The gold fields and Canterbury', *Press*, 3 August 1861, 2.

55 *Lyttelton Times*, 17 August 1861, 6.

56 *Lyttelton Times*, 21 May 1859, 4, 7.

57 'Obituary: William Smart', *Star*, 12 October 1900, 3.

58 *Lyttelton Times*, 14 September 1861, 4.

59 *Lyttelton Times*, 5 February 1853, 7; 19 March 1853, 12.

60 'Provincial Council', *Lyttelton Times*, 18 April 1857, 5; *Lyttelton Times*, 25 April 1857, 6.

61 *Otago Witness*, 29 June 1861, 5; Otago Provincial Superintendent, Message No. 4, 'Gold discovery bonus', 5 November 1861, AAAC 701 D500/8e 55, ANZ, Dunedin.

62 *Lyttelton Times*, 10 August 1861, 4.

63 *Lyttelton Times*, 5 October 1861, 4.

64 Ibid.

65 *Lyttelton Times*, 15 January 1862, 4; Julius Haast, 'Notes on the geology of Canterbury', *Lyttelton Times*, 12 November 1862, 3.

66 *Press*, 24 May 1862, 12.

67 *Lyttelton Times*, 30 August 1862, 4.

68 *Otago Witness*, 23 September 1862, 4.

69 *Lyttelton Times*, 3 September 1862, 4

70 Ibid., 5.

71 Herbert Charlton Howitt (1838–1863) was a younger son of William and Mary Howitt, well-known English Quaker writers. In 1856, aged 18, he emigrated to Australia with his father and brother, and continued on to New Zealand in 1862. In June 1863 he and two members of his track-cutting party drowned in Lake Brunner while out in his canoe catching eels. Charlton Howitt, 'Diary of Mr. Howitt during his late search for gold, 12 September–30 November 1862', qMS-1015, ATL; 'Bridle track to the Grey', *Lyttelton Times*, 10 January 1863, 3.

72 Charlton Howitt to Provincial Secretary, 2 September 1862, ICPS 1375/1862, ANZ, Christchurch.

73 According to Arthur Dobson, Māori (Ngāi Tahu and Poutini Ngāi Tahu) had long used the route across the Taramakau Saddle between east and west, but they always went through the bush and along the riverbed; they did not make a track, so as to prevent their enemies from following them and to protect their pounamu/greenstone sources on the Coast. Arthur Dudley Dobson, 'The opening of the Canterbury West Coast', in Robert Allan (ed.), *Reminiscences: Early Christchurch and Canterbury*, 1922, christchurchcitylibraries.com/DigitalCollection/Archives/Archive210/PDF/Reminiscences-1922-Dobson.pdf.

74 Julius Haast, *Geology of the Provinces of Canterbury and Westland, New Zealand*, http://nzetc.victoria.ac.nz//tm/scholarly/tei-HaaGeol-t1-body1-d1-d1-d6.html; *Lyttelton Times*, 6 September 1862, 4; *Otago Daily Times*, 22 September 1862, 4; *Lyttelton Times*, 4 October 1862, 4.

75 *Lyttelton Times*, 22 October 1862, 2.

76 *Lyttelton Times*, 13 September 1862, 4.

77 'William Smart: Obituary', *Star*, 12 October 1900, 3.

78 'Public works', *Lyttelton Times*, 10 January 1863, 3.

79 William Martin, 'A pioneer's reminiscences', [c. 1913], MS206, Hocken Library, University of Otago, pp. 1, 6–11; *Otago Daily Times*, 13 October 1862, 4; *Otago Witness*, 4 November 1862, 4; *Wellington Independent*, 6 November 1862, 3; *Colonist*, 28 November 1862, 2.

80 John P. Oakes, 'Discovery of gold on the West Coast: To the editor of the *Lyttelton Times*', *Lyttelton Times*, 21 March 1865, 5; Martin, 'A pioneer's reminiscences', pp. 2–4.

81 *Lyttelton Times*, 20 December 1862, 4.

82 Martin in Bradshaw (ed.), *The Diggers' Story*, p. 37. Like Smart, Martin was one of the earliest Pākehā prospectors in West Canterbury. He arrived on the Coast on Dixon's return voyage in mid 1863, and worked until the end of that year at the Hōhonu diggings with Māori miners Ihaia Tainui, Himona ('Simon') Tuakau and Iwikau te Aika ('Samuel'). After leaving the Coast he remained in touch with his Māori friends, who would send him news from the diggings.

83 'Shotty' gold: small, granular pieces, resembling shotgun pellets.

84 Martin, 'A pioneer's reminiscences', pp. 4–6; *Lyttelton Times*, 20 December 1862, 4.

85 *Lyttelton Times*, 20 December 1862, 4.

86 Richard Sherrin (1832–1893), who came to Canterbury via the New South Wales and Victorian goldfields, was a member of Drake's West Coast survey expedition in early 1863, and wrote a secret report for the Canterbury government on the Coast's prospects; his expedition journal was serialised in the *Press*. He returned to the Coast later that year and was engaged in gold digging at the Hōhonu; he explored the coast to the south, helped search for Howitt's body, and finished cutting his track. He became a journalist and wrote several books, including *Early History of New Zealand from Earliest Times to 1840* (1890).

87 Thomas Dixon, 'Letter to John Oakes', 7 December 1862, *Lyttelton Times*, 20 December 1862, 4; 'The West Coast: Mr. Sherrin's journal', *Press*, 17 December 1863, 2.

88 'Bridle track to the Grey', *Lyttelton Times*, 10 January 1863, 3.

89 May, 'West Canterbury gold rush to 1865', p. 53; *Lyttelton Times*, 24 December 1862, 4.

90 In those days the Hōhonu flowed directly into the Taramakau about 10 kilometres from the river mouth. Today it meanders parallel to it for some kilometres and disappears into the shingle about 4 kilometres from the beach.

91 May, *The West Coast Gold Rushes*, p. 73.

92 Charles Money, *Knocking about in New Zealand*, Capper Press, Christchurch 1871/1972, p. 47.

93 *Colonist*, 27 February 1863, 3; R. A. A. Sherrin, 'Journal of an expedition to the West Coast', *Press*, 24 March 1863, 2.

94 'West Coast diggings', *Colonist*, 27 February 1863, 3; *Lyttelton Times*, 28 February 1863, 4; Charlton Howitt, 'Journal of H. C. Howitt, 1863', MS-1038, p. 22, ATL.

95 Waite, *Narrative of the Discovery*, p. 9; R. A. A. Sherrin, 'Journal of an expedition to the gold field of the Taramakau', *Press*, 25 March 1863, 1.

96 Hilary Low, *Pushing his Luck: Report of the expedition and death of Henry Whitcombe, by Jakob Lauper*, Canterbury University Press, Christchurch, 2010, pp. 41–44; R. A. A. Sherrin, 'The West Coast: Mr. Sherrin's journal (cont)', *Press*, 17 December 1863, 2.

97 'The early 60s: Food for the diggers', *Press*, 10 February 1928, 4.

98 Martin, 'A pioneer's reminiscences', p. 3.

99 'The West Coast', *Press*, 24 March 1863, 1.

# Part 2

## The Westland Goldfields, from the diary kept by Wm Smart, one of the first discoverers of payable gold in that part of N. Zealand

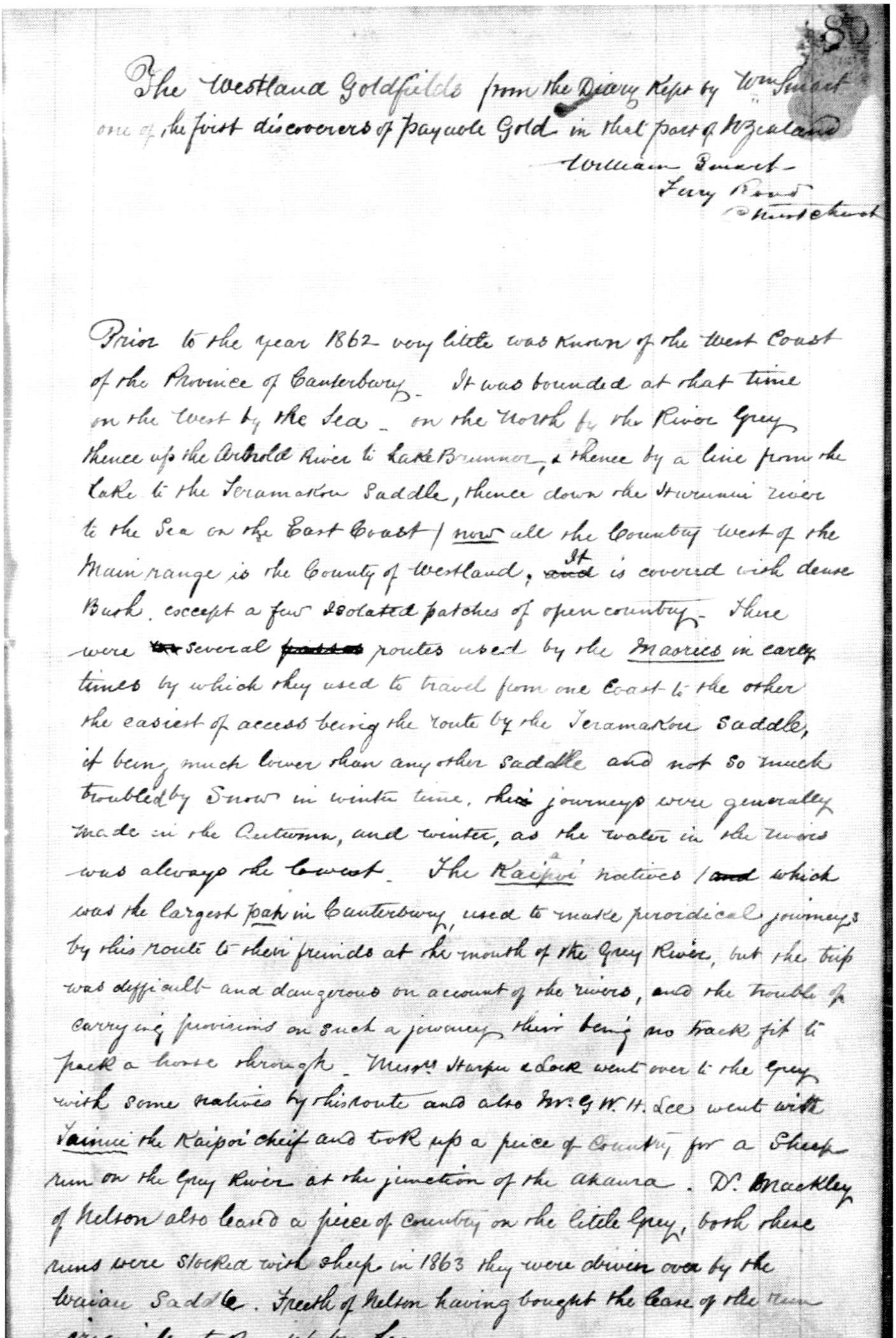
The Westland Goldfields from the Diary Kept by Wm Smart
one of the first discoverers of payable Gold in that part of N Zealand
William Smart
Fery Road
Christchurch

Prior to the year 1862 very little was known of the West Coast of the Province of Canterbury. It was bounded at that time on the West by the Sea – on the North by the River Grey thence up the Arnold River to Lake Brunner, & thence by a line from the lake to the Teramakau Saddle, thence down the Hurunui river to the Sea on the East Coast / now all the Country West of the Main range is the County of Westland, ~~and~~ It is covered with dense Bush except a few isolated patches of open country. There were several ~~passes~~ routes used by the Maories in early times by which they used to travel from one Coast to the other the easiest of access being the route by the Teramakau Saddle, it being much lower than any other saddle and not so much troubled by Snow in winter time, the journeys were generally made in the Autumn, and winter, as the water in the rivers was always the lowest. The Kaiapoi natives (and which was the largest pah in Canterbury, used to make periodical journeys by this route to their friends at the mouth of the Grey River, but the trip was difficult and dangerous on account of the rivers, and the trouble of carrying provisions on such a journey there being no track fit to pack a horse through. Messrs Harper & Lock went over to the Grey with some natives by this route and also Mr G W H Lee went with Tainui the Kaiapoi chief and took up a piece of Country for a Sheep run on the Grey River at the junction of the Ahaura. Dr Brackley of Nelson also leased a piece of country on the little Grey, both these runs were stocked with sheep in 1863 they were driven over by the Waiau Saddle. Freeth of Nelson having bought the lease of the run originally taken up by Lee

This is the first page of Smart's manuscript. He wrote in the unused part of an old ledger book with large numbers printed at the top of each page, filling pages 79 to 176.

## Transcribing and editing notes

'The Westland Goldfields, from the diary kept by Wm Smart …', written between 1887 and 1890, is the longest and most detailed of a number of accounts, both published and unpublished, that William Smart wrote of his time on the West Coast, all based on his diary. The first of these, 'Diary of a prospecting tour to the West Coast', he wrote soon after the events took place, as part of his 1865 application for the Canterbury government's gold reward. This earlier text tells part of the story in more detail, some of which I have incorporated into my transcription of 'The Westland Goldfields'.

Smart's sentences are often long and complex, and where this seems excessive I have split them into shorter sentences. Punctuation has been added where needed to aid the reader, and the capitalisation of ordinary nouns has been changed to lower case. His many abbreviations have been written out in full; for example, 'Chch' (Christchurch), 'f. pan' (frying pan) and 'Secty' (Secretary).

Most of Smart's factual errors have been corrected silently (i.e., without editorial comment); more significant ones have been corrected either in square brackets or in a footnote. Occasionally a word has been inserted where Smart obviously omitted one, or to clarify an otherwise obscure sentence. Most of his Victorian spellings and usages have been retained, such as 'shewed' for showed, 'pah' for pā, and 'Maories' for Māori.

Smart's manuscript shows his own editing process, with additions or comments he inserted later or as marginalia. These have been incorporated into the text. Where his additions are repetitious, they have been edited. My own insertions appear in the text in square brackets; and fuller explanations and definitions are in the footnotes.

Many people appear in this story. Smart identifies some of them, but the information is uneven. Where possible, I have provided brief details on the main players. First names and a few key dates have been added silently; and more detailed information can be found in square brackets or in footnotes. For more on these people, and on others who appear in the text, see *Te Ara – the Encyclopedia of New Zealand*, www.teara.govt.nz; Philip May, *The West Coast Gold Rushes*, 1962; and Mark Pickering, *The Colours*, 2011.

Smart's pen-and-wash drawings that accompanied 'The Westland Goldfields' are reproduced throughout the text.

***Wharfedale Country***
*Country prospected for Gold previous to going over to West Coast. This Country was taken up first by Lee and Ellis. A track was cut to it through from Birch Hill over the Birch Range – a large valley about 4 miles wide the upper part rich swamp land. The Ashley gorge cuts through some good slate country in the first gorge, then divides the plain in half and through a 2nd Gorge to the Pakiteraki [Puketeraki] Range.*

'Lee' was Smart's friend Jockey Lee, whose sheep run, 'Wharfedale', took in all the flat country in the Lees Valley, which Smart has labelled 'Waiotite[?] Plain', bounded by the Ashley and Whistler rivers. Above the line of the hills he has written 'Ashley Gorge', 'R Townsend' and 'Pakiteraki' [Puketeraki Range].

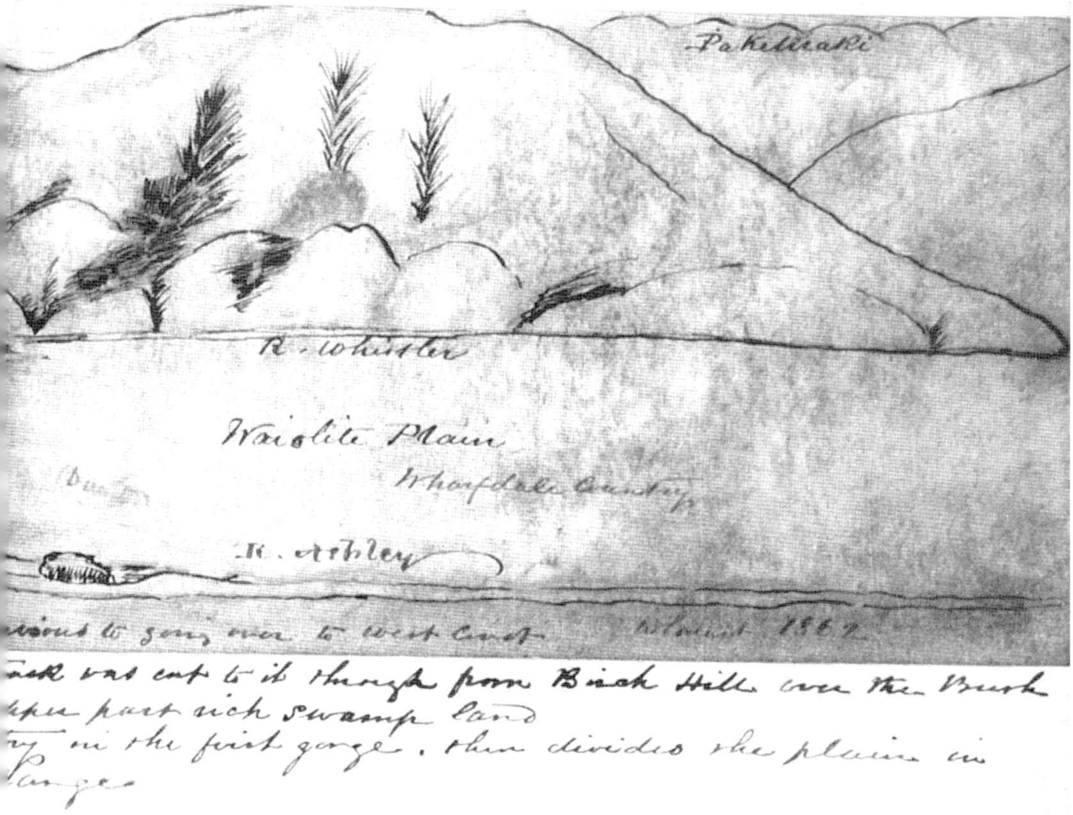


## The Westland Goldfields, from the diary kept by Wm Smart ...

Prior to the year 1862, very little was known of the West Coast of the Province of Canterbury. It was bounded at that time on the west by the sea, on the north by the River Grey,[1] thence up the Arnold River to Lake Brunner, and thence by a line from the lake to the Teramakau Saddle,[2] thence down the Hurunui River to the sea on the eastern coast. Now all the country west of the main range is the County of Westland. It is covered with dense bush except for a few isolated patches of open country.

There were several routes used by the Maories in early times by which they used to travel from one coast to the other; the easiest of access being

1 Māori called this the Māwhera; it was renamed the Grey River by Thomas Brunner.

2 The saddle was known by several names: the Taramakau Saddle (various spellings), the Hurunui Saddle and, officially since 1857, Harper Pass – but this name was seldom used in Smart's time.

***Pakiteraki [Puketeraki] Range. W. Smart 1862***
*Prospected this country for Gold previous to going to West Coast. Plenty of snow on this range while we were here, we were stuck up on top of the range by heavy snow for 3 days.*

To the right of 'Mt Pember', Smart has annotated the line of a trail: 'Track up to the valley of the Esk due north on top of the Snowy Range'.

the route by the Teramakau Saddle, it being much lower than any other saddle and not so much troubled by snow in winter time. The journeys were generally made in the autumn and winter, as the water in the rivers was always the lowest. The natives of Kaiapoi, which was the largest pah in Canterbury, used to make periodical journeys by this route to their friends at the mouth of the Grey River, but the trip was difficult and dangerous on account of the rivers and the trouble of carrying provisions on such a journey, there being no track fit to pack a horse through.

Messrs Leonard Harper and Edwin Lock went over to the Grey with some natives by this route in 1857, and also Mr. G. W. H. ['Jockey'] Lee went with Werita [Wereta] Tainui, the Kaiapoi chief, in 1858, and took up a piece of country for a sheep run on the Grey River at the junction of the Ahaura, about 20 miles from the coast. Dr. Mackley of Nelson also leased a piece of country on the Little Grey [the Māwhera-iti]; both these runs were stocked with sheep in 1863, which were driven over by the Waiau Saddle, Isaac Freeth of Nelson having bought the lease of the run originally taken up by Lee.

The Nelson portion of the West Coast, which joins the Canterbury boundary

***Gorge of Fergusson's Creek falling into River Townsend [Townshend]. W. Smart 1862***

Another scene Smart drew on his prospecting expedition in the North Canterbury hills. The Townshend River flows into the Ashley.

on the north [side of the Grey], had been more explored and partly surveyed. Mr. Brunner, the Chief Surveyor, had been on a trip which occupied about two years. He went from Nelson as far south as Lake Brunner; also Messrs James, James Jr, and Alexander Mackay and some others.

In 1859 Dr. Haast, who was then the Geologist for the Province of Nelson, had been sent there also to geologically examine the country; but singular to say, this report was very unfavourable to the existence of gold in that country. Part of his report winds up as follows:

> North of the Buller, in the Maruia, as well as in the whole course of the Grey and its tributaries, rarely leaving untried any spot which seemed likely, we searched in vain, unable to detect the least sign of the precious metal.[1]

The Doctor was afterwards Geologist for Canterbury, and when we sent over some gold by a Maori to Christchurch in January 1863 in order to get a schooner sent round with stores, the Doctor's report prevented our getting the assistance we required, and after twelve months' delay we had to open the goldfield with the assistance of Nelson storekeepers.

1 Julius Haast, *Report of a Topographical and Geological Exploration of the Western Districts of the Nelson Province, New Zealand*, Nelson Provincial Government, Nelson, 1861, p. 122.

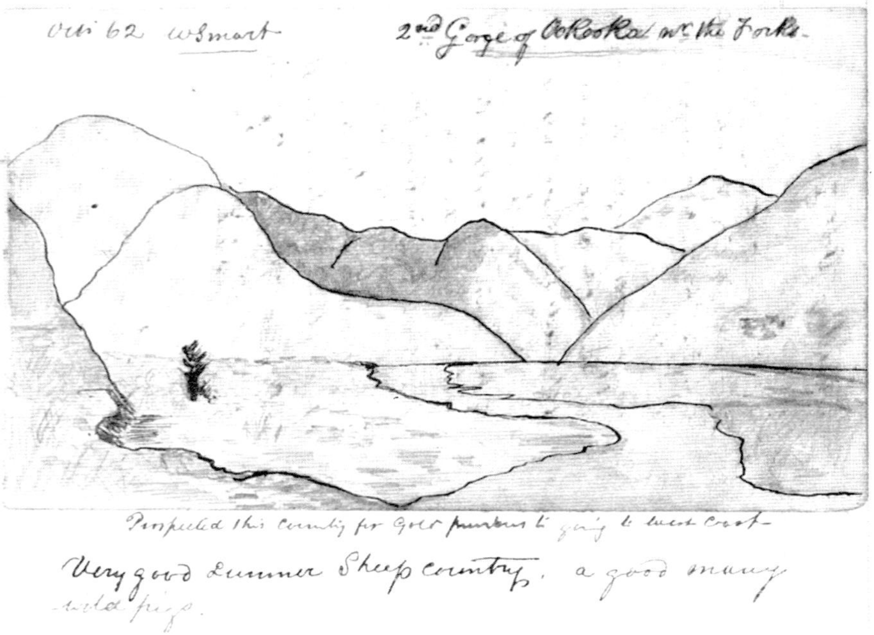


***2nd Gorge of Ookooka [Okuku River] near the Forks. Oct '62 W. Smart***
*Prospected this country for Gold prior to going to West Coast. Very good summer Sheep country, a good many wild pigs.*

Here Smart's handwriting shows through from the other side of the page of his original diary.

In late 1862, the *Emerald Isle*'s Captain Oakes went to the Grey with stores for Freeth's sheep run. This vessel went into the Teramakau in mistake for the Grey, got ashore [ran aground] and the Maories assisted to get her off. The stores were left in the charge of Terapuhi [Tarapuhi], the Grey chief, until such time as Freeth should arrive overland from Nelson. These stores kept some of the earliest arrivals previous to the outbreak of a goldfield from actual starvation, as the old chief sold a few pounds of flour to one and another, but only sufficient to carry them back to Taylor's station[1] on the Hurunui, or to the Buller River where there were, at that time, the only storekeepers on the Coast.

In 1860, before these stores were started, Messrs James, Charles and Ambrose Jackson, James Smart [no relation to William], William Belcher and others went to the Buller River in the cutter *Fantome* on a prospecting expedition. They made a canoe, planted some garden seeds, prospected about

1 Until the Arthur's Pass road opened in 1865, Henry Taylor's sheep run, 'The Lakes', east of the Taramakau Saddle, was an important stopping place for early foot travellers to and from the West Coast goldfields. Taylor sold them provisions and offered shelter in his woolshed. L. D. G. Acland, *Early Canterbury Runs*, revised edn, Whitcombe & Tombs, Christchurch, 1946, pp. 284–85.

the Buller River and fell in with [joined] Haast's party, and some of them returned overland with him to Nelson.

Some Collingwood miners and some Maories were the first to find gold in the Buller district at the Waimangaroa Creek, about nine miles north of the Buller.[1] There was some coarse nuggetty gold got in this creek, but it was very rough, heavy working amongst large boulders. The creek was nearly deserted at Christmas 1862 by the find of gold at the Lyell Creek, which is a tributary of the Buller, about 40 miles by the river up from its mouth.

The schooner *Ada* from Canterbury came into the Buller in August 1862. She was bound down south on an exploring trip, but unfortunately one of her party, Mr. C. M. Ollivier, died on board and was buried at Jackson's Bay.

In 1861[2] the Canterbury Provincial Government offered a reward of £1000 for a payable goldfield, and myself, Michael French,[3] Cook, Ferguson, Robert Day and David Everest went to the back ranges with pack horses etc, and were prospecting from September to December 1862, but could find no gold on this eastern watershed of the ranges.

1862: In December 1862, myself and Robert Day interviewed Mr. Maude, the Provincial Secretary, in Christchurch, and learnt the terms on which the reward for discovery of a goldfield would be given. We found that the Provincial Government had sent Mr. Howitt and a party to cut a track by the Teramakau Saddle route to the West Coast beach, and the government had decided to have a survey made of that portion of Canterbury.[4] It was stipulated with us that if Howitt's party found payable gold, it would not invalidate our claim to the reward were we successful in also finding payable gold.

Day, Fraser and myself therefore agreed to start at once, after a long interview with Tainui and some more Maories at Kaiapoi,[5] who described to us, by drawings on the sand, the route to follow and which I took particulars of. We

---

1 These were not the earliest diggings in the district; the first were 'the Old Diggings' up the Buller River.

2 Smart wrote 1862 in error.

3 Michael French, a native of County Cork, was born around 1822 into a family of landed gentry. He was said to have been an officer in the East India Company army. The 1860 Christchurch militia list cites him as Michael Philip French, gentleman of Kaiapoi, which is where he would have met Smart.

4 This was Drake's survey expedition, which took place from January to March 1863.

5 This 'Tainui' was either Kaiapoi chief Wereta Tainui, or his son Ihaia Tainui who, at about that time, had passed through Kaiapoi en route to Christchurch carrying his and Captain Dixon's gold and their applications for the government's gold reward.

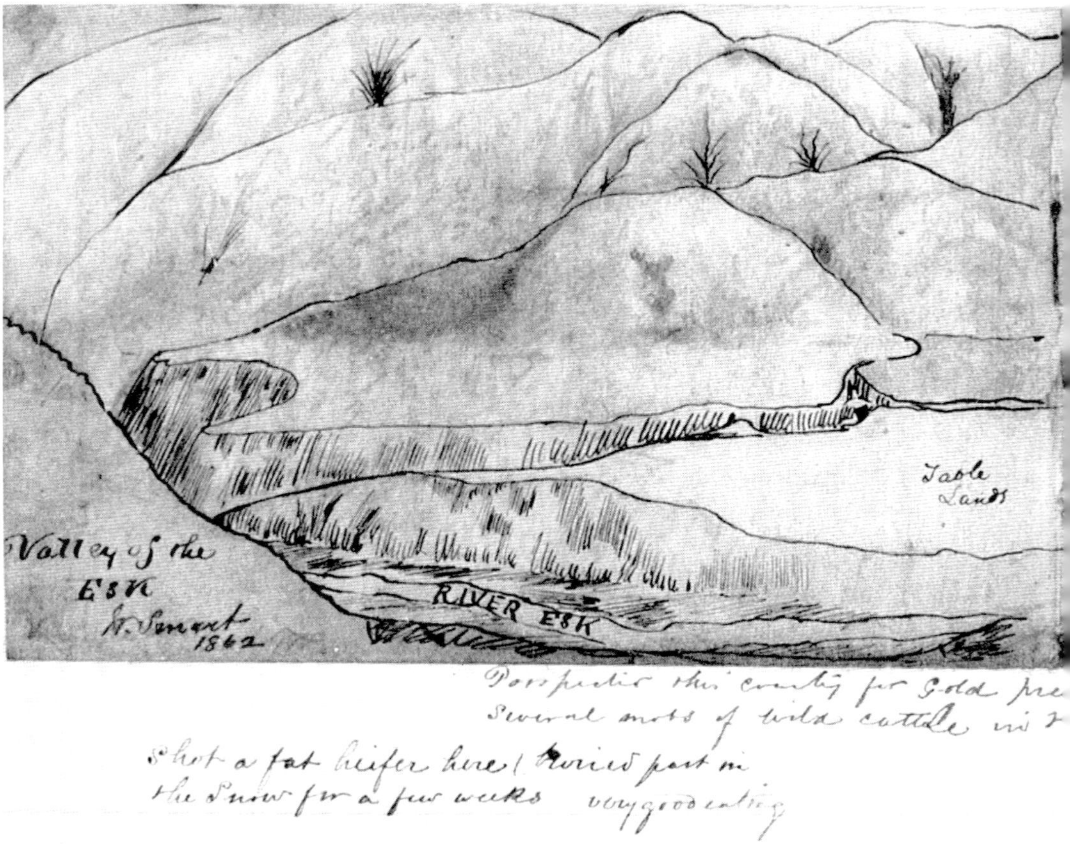


***Valley of the Esk. W. Smart 1862***
*Prospected this country for Gold previous to going to the West Coast 1862. Several mobs of wild cattle in this country (afterwards called Lochinvar [station]). Shot a fat heifer here (buried part in the snow for a few weeks). Very good eating.*

reached Mallock and Lance's station at Horsley Downs[1] on 24 December 1862 with pack horses and tents, tools and provisions. We here got a 200lb bag of flour and went on to Taylor's woolshed in the Waitohi and camped for the night.

1862, December 25, 26: Left the woolshed and crossed the south branch of the Hurunui, and as it commenced to rain we camped early. Called at Taylor's station and had some dinner, and this being the last house on the road [before the Taramakau Saddle], we overhauled our swags to see that we had everything necessary. Taylor told us that Howitt's party had returned to

1 Horsley Downs was an early North Canterbury sheep run. Like Taylor's it was a stopping place for foot travellers en route to the West Coast via the Taramakau Saddle, until the Arthur's Pass road opened.

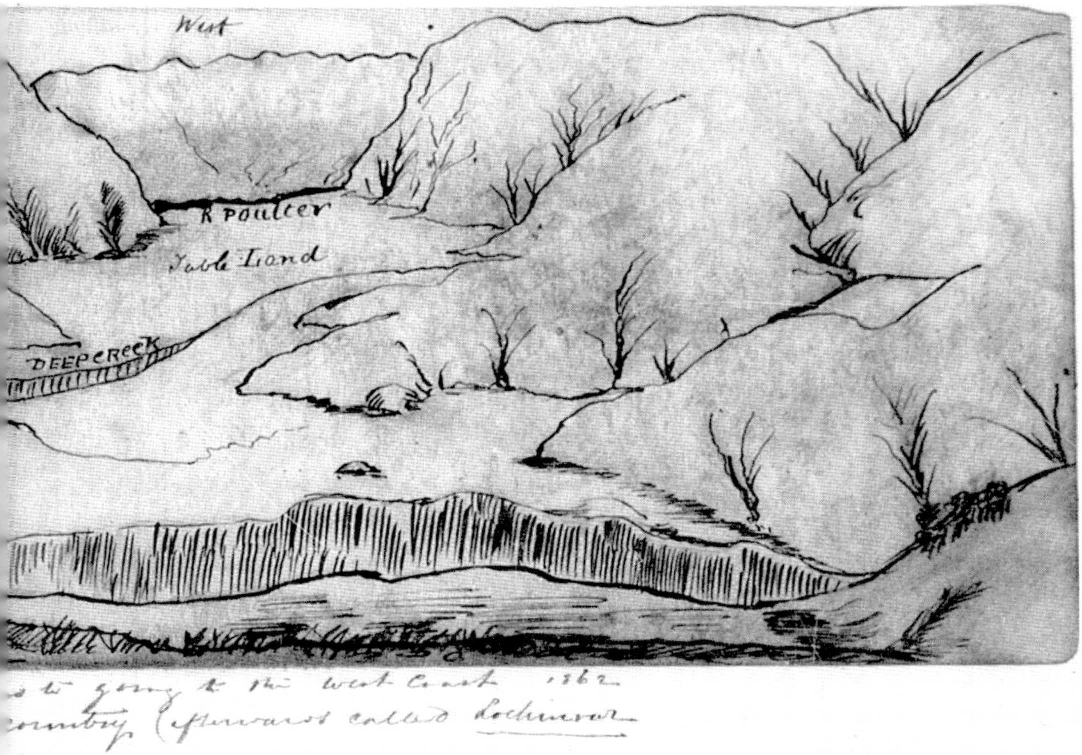


Christchurch some days previous for stores and to stay over Christmas, and that they were only just on the Saddle with the track they were cutting.

We crossed round by the lakes – in some places very boggy – and pitched our camp at the head of Lake Sumner. Caught a large silver eel in the lake which was splendid eating.

This is very nice country, good land, plenty of bush on the Nelson side[1] of the lake, which is about 12 miles by four miles. A good place for an inland township – plenty of water for mill purposes and easy to make a good road to it. It's a wide valley between the mountains, with the Hurunui River running into the lake and out of it again through the gorge into the plains. The Maories called this upper portion of the Hurunui, from the mountains to the lake, the Eastern Teramakau. The smaller lakes we passed round are Lakes Taylor, Catherine and Mary.

1 The lake and the Hurunui River form the eastern boundary between Canterbury and Nelson provinces.

***Under Mt Pember. Wm Smart, Oct 1862***
*In the early morning. Prospected this Country for Gold previous to going to West Coast.*

***River Esk running South. WS 1862***
*Very high shingle Terraces, grass, and bush.*

Rain pretty heavy during the night.

Dec 27: Made a late start on account of the rain. At the head of the lake saw two men at a grass hut. These we found were Messrs [Rowland] Davis and [Charles] Money, who had been prospecting in the Virginia country[1] for gold, but without success. As they were making a flying trip to the Saddle to look at the country, they had only a blanket and a very few provisions with them. They walked on ahead of us, as they meant to return for stores after looking at the Saddle. We went on a few miles and then camped, as it came on to rain heavily.

Dec 28, 29: Camped, raining very hard and river rising. Plenty of feed for the horses, plenty of firewood. Shot four ducks.

Dec 30: Fine. Crossed the river once, but rather deep, so thought it advisable to camp where there was plenty of feed for the horses, but the Saddle was no great distance away.

There is a hot water spring on a spur of the range on the right-hand side of the river, about 4 miles east of the Saddle. The steam can be seen rising from it in the early morning.

Dec 31: Followed up the river all day, crossing many times. In the last crossing one of our horses fell into a hole onto his back and wetted some of our swag, spoiling a greater portion of our sugar and matches. Camped on top of the Saddle.

Here the Teramakau running west and the Hurunui running east rise within a few yards of each other; the eastern side having fine open grassed flats all the way up, and a dray and bullocks can be driven from Taylor's station to the top of the Saddle with perfect ease – not so bad as the road over the Weka Pass used to be a few years ago – and we were all much surprised at its easy access.

But looking westward there is nothing to be seen but dense bush, thick undergrowth and shingle bed running down to a black-looking gorge. No chance of taking horses any further as there is no track or feed, so we shall have some heavy work to carry all our stores from here westward. The top of the Saddle very swampy.

1863 Jan 1st: We turned our horses adrift. Stuck our saddles in a tree on top of the Saddle, and portioned out our swags, which were very heavy, having brought two hundred pounds of flour besides other necessaries.

We started into the bush track down the western side, following a water

1 'Virginia country' was a district in the North Canterbury hills on the Waitohi River.

***The River of the Grey Mountain. Poulter. 1862 WS***
*Gorge of a Branch of Waimakariri River looking due North. Prospecting for Gold '62 before going [to the] West Coast.*

***Branch of Waimakariri River looking due South. 1862 W. Smart***

course between the two branches of the Teramakau, to 'Howitt's' last camp, about 1½ miles down from the Saddle. Camped there for the night as it was unoccupied, Mr. Howitt and his party having gone to Christchurch. Shot several pigeons close to the camp.

To our surprise at dark who should come in but Davis and Money, bringing with them a Frenchman and a German[1] who they had found a few miles down the Teramakau in a state of starvation. And most fortunate it was for them that we were so near, for Davis and Money's provisions were all gone, but the German and Frenchman had had nothing to eat save a raw wood-hen [weka] for several days.

They had been nearly drowned in the river during the last rain, wetted all their matches, and were in a most miserable plight. We fed them all night on stewed pigeon and damper. We found that these men had come on Howitt's track and gone down [to the West Coast] prospecting, and Money and Davis fell in with them and brought them back.

1863 Jan 2: Left the men enough provisions to take them back to Taylor's station and started on down the river. Very rough travelling with our heavy swags through scrub, bush, and boulders. We crossed the river 14 times this day, and camped late. A very hard day's work. S.W. by W.

We carried a tent, two flys, our own clothing and blankets, pick, shovel, tin dish, frying pan, gun, 14lbs of shot, 2lbs of powder, caps, bacon, tea, sugar, 200lbs of flour, salt, matches, candles, a few nails – nearly 1 cwt [hundredweight] each.

Jan 3: Crossed the river four times and camped early. The last indications here of Howitt's party: some scrub recently cut and a piece of red rag on a stick. Named this Maori Camp.

Jan 4: Crossed the river and walked down until we came to a good-sized creek coming in from the south side, called by the Maories the Otiaki [Otehake]. This creek rises from a saddle (since called Worsley's Pass); the eastern watershed is a branch of the Waimakariri, called the Poulter.

The travelling was very rough as we had to follow down the shingle beds – or rather boulder beds – until we were again compelled to cross the streams where the spurs from the ranges come down to the river in precipitous bluffs. All dense bush, thick undergrowth of vines and supplejack, tutu trees[2] a foot in diameter, mosses and ferns of great variety. Slate country, and quartz in places.

---

1 The pair were Jakob Lauper (a Swiss, identified here as the Frenchman) and 'Sam Johnson' (the German).

2 Tutu is a native shrub: all parts except the flower petals are poisonous.

***Maori Camp about 13 miles West of Teramakau Saddle. Jan'y 3 1863, W. Smart***
*Copy from Diary*

Jan 5: Prospecting: got a few colours[1] and black iron sand. Began to rain in the afternoon. Four men came down; one of them, a darky, Arthur,[2] had been a cook at one of the hotels in Christchurch. They had come over prospecting, but brought very little food with them and seemed to have no idea that they were going to a place where there was no food to be got but what they could catch or shoot.

Jan 6: Raining. All camped. Prospecting. Plenty of firewood and wood-hens which we caught with a snare, to save powder and shot.

Jan 7: Fine, but plenty of water in the river. Walked about 6 miles down the river cutting our way through bush to avoid the crossings, which are very much flooded. Easier travelling today as we passed through several patches of open ground on the banks of the river. Camped early. Prospecting.

Jan 8: In crossing the river this morning, one of our party was capsized, but his swag kept him up and we got him out a little lower downstream. He was all right with the exception of a ducking and a few pounds extra weight to carry in his swag.

Cut our way through bush which was anything but pleasant – supplejack

1 Colours: gold specks in a prospector's pan, or in a sample of gravel or quartz.

2 Arthur Addison, an African-American who was said to have been a freed slave.

catching your swag behind and bush lawyer[1] taking your hat off, and otherwise damaging your physiognomy, made us say anything but our prayers. The river running S.W. by W. average bearings.

Crossed another larger stream coming in from the south side. There was a heavy freshet[2] in it, and we had to go up the river over a mile and cross it in three streams,[3] which we managed with difficulty, it being so rapid.

This is called the Otira and also rises from a saddle, the western watershed of which is another branch of the Waimakariri [the Bealey River]. The Maoris had a pass by this route but it was very rough; their old track came through the Woodstock station beyond Oxford.

Jan 9: Camped. Heavy rain all day. Plenty of kakas – very large. Tutu. The boulders not so large but still very rough walking. All the hills have a rounded appearance in short spurs running down from the main ranges which run at right angles, all covered with dense bush. Very few bare spots to be seen and they being shingle slips caused by melting of the snow. The shingle a hard slate drift[4] in some places.

Jan 10: Fine, but the river so high that we started to cut our way through the bush; very interlaced with supplejack and bush lawyer. All day very bad travelling – slate drifts, boulders and quartz.

Towards evening found we should be obliged to cross on account of a steep bluff, so we all crossed together in a very deep ford, but we kept well above the rapid to avoid the strong current. Camped late.

Jan 11: Cutting our way all day through bush on the northern side of the river to avoid crossing as the river is very high. The mountains here break into a flat which forms the open country to Lake Brunner and by going through about a mile of bush there is some grassed country of about 4000 acres in extent.[5] But compelled to camp early as we came to a steep bluff, the Inchbonnie Bluff, of granite rock. This is the first granite we have seen and is just above a large river coming in from the south, which will be the Opiaukura [Hopeakoa] or Taipo River.

---

1 Bush lawyer (tātaramoa): a plant with hooked thorns that snag clothing and rip the skin. It gained its English name because once it becomes attached to a victim it will not let go until it has drawn blood.

2 A fresh or freshet: a rush or sudden flood of water in a river or creek after rain.

3 To cross in three streams: to follow a river upstream and cross it in stages by way of its smaller tributaries.

4 Drift: rock debris piled up in heaps by the action of water and/or glaciers.

5 This was known as 'the Pakihi' and was where Howitt grazed his horses. The word pākihi means open country, grassland or barren land.

Jan 12: The men who have been travelling with us left us this morning, intending to cut their way over the bluff, but we intend to find a ford and cross the river if possible. We had just packed up our swags and reached the river bank, and were on the point of starting when we saw some Maories come out of the bush on the opposite side of the river. We cooee'd[1] to them and found they were Jack [Ihaia] Tainui, Patterson, Peter Mutu and Harry from Kaiapoi.[2] After a korero [discussion] across the river, we found that to get a ford we should have to retrace our steps about three miles, so it was agreed that they should cut a long sapling and cross over to us and help us over. This was done and we recrossed with them in very deep water.

The tallest and strongest Maori took one end of the pole up stream, another the end downstream. We then stationed ourselves at equal distances, taking a firm hold of the pole and pressing upstream. We then walked in the river all abreast and in a slightly slanting direction across the river. The upper man broke the force of the current, which was flowing over his shoulder, and we came over very easily, our swags of course getting very wet. By this means I have known natives to cross rapid deep rivers which otherwise would be impossible.

A few hundred yards brought us to the Taipo River, a stream as large as the Teramakau at its junction, coming in from the southwards, which we all crossed in the same manner. After crossing we wrung out our blankets, had some dinner and re-adjusted our swags, the Maories helping to carry part of ours.

We got a colour or two of gold in the loose shingle in the river bank at the mouth of the Taipo – very fine. This is a very dangerous, rapid river, full of large boulders, and very bad to cross. It is the terror of the natives in coming over, as it is subject to floods more than the upper part of the Teramakau. They say it is the only one that stops them in a journey. They therefore re-christened it the Taipo.[3]

After about five miles walk further down the Teramakau riverbed and crossing one or two branches, we came to an island covered with flax, where

---

1 'Cooee': a call to attract attention, adopted by Australian and New Zealand settlers from the Aboriginal Dharuk language, 'guwi'.

2 John Patterson was the English name of Hone Paratene Tamanui-a-Rangi, a former whaler, and one of the early miners at Greenstone Creek. In 1868 he became the first MP for the Southern Māori electorate. Peter Mutu was the official mail carrier between Christchurch and the Grey in 1864, and was for many years a butcher at 'Purau', the Rhodes family sheep run on Banks Peninsula. He was father of the well-known minister, the Rev. George Peter Mutu. Harry was probably Henare Mahuika, later a Native Constable. *Press*, 27 June 1879, 3.

3 'Taipo' is likely to be an English corruption of tipua or tupua (a goblin or demon), a word that Māori associated with this treacherous river.

we camped. We pitched the tents while the Maories started to get flax sticks to make a mogie[1] to carry us down the river for the rest of the distance.

The Teramakau takes here a bend to the North, opening out into bush and a very large river flat some miles in width and called by the Maories the 'Perou Nui'.[2] From here we can get a splendid view of all the surrounding mountains.

The Maories told me here the names of the different mountains and streams we had crossed and all the places they had been to hunt for kiwis and other birds, the position of Lake Brunner, and to the South West a low opening running through towards the beach by the river Arahura. The Wainoni Creek rising in a very low terrace or saddle and running to the Teramakau and another creek, also called Wainoni, running from that saddle into the 'Arahura River'.

All these Maories had been over here from Kaiapoi several times and lived at the Grey when they were children; their fathers or friends took them over when the Kaiapoi pah was being besieged by Te Rauparaha and other North Islanders.[3] One mountain they shewed me as being on the Arahura, which they said contained a great quantity of greenstone, but was very difficult of access.

It was beautiful, hot, sunny weather and we got all our blankets, etc, well dried before sundown. Saw shoals of fish coming up the river – larger than mullet.[4]

Jan 13: Up at daybreak and finished the mogie by midday. It is made with dry flax sticks as follows. Two large rolls of flax sticks are bound together in the shape of two huge cigars, small at each end and about three feet in diameter in the middle and about 25 feet long. The ends of each of the rolls are then tied together, the centre is spread out by pieces of dry wood to about six feet wide, and the centre is then filled up with more dry flax sticks and strongly laced with flax lashings. Small bundles are tied on top to form a gunwale on the top sides.

Putting all our swags on these, and with some manuka poles and flax paddles, we made a start downstream. It carried eight of us with swags about a foot above the water. We drifted down, over several shallow falls [rapids]

1 Mogie: an English variation of the Māori word mōkihi or mōkī, a temporary raft used on lakes, estuaries or rivers. Constructed of bundles of reeds, flax stalks or rushes tied together, they were quick to make using materials to hand. See: www.teara.govt.nz/en/object/6015/mokihi.

2 This was the location of the Taramakau settlement, at the foot of Mt Smart. The name 'Perounui' has disappeared from current maps.

3 The siege of Kaiapoi occurred in 1831.

4 Probably the upokororo or New Zealand grayling, once prolific but now extinct.

until we came to the gorge. Here the formation changes to blue marl[1] bluffs, with red and blue granite and slate wash. On one of these marl banks in the gorge we landed. (It was a little distance up the river, above where the Teramakau Bridge now [1890] crosses from Kumara to the Greenstone.) Day picked up a piece of gold as large as a shilling. This was the first piece of gold got in this part of the country.[2]

We started again and in a short distance came to a good-sized creek coming in on the north side of the river. This is the largest stream we have met with on that side. Here we saw a similar reef and wash, and agreed to camp and prospect for gold, as being the best-looking spot we have seen on the river. This was called the Little Ohono Creek,[3] which I consider about 50 miles from the Saddle.

The Ohono has a clay slate bottom, loose shingle drift, granite boulders. The country is low terrace land, intersected by deep creeks and clay slate reefs rising in many places to 40 and 50 feet. All the country covered with thick bush. A great many varieties of ferns, and most parts of the bush very boggy. Trees smothered with parasites.[4]

We tried the banks of the creek and got a few colours in nearly every dish. We then arranged with the Maories to go mates with us,[5] and started the Maories on to the Grey to bring back with them a pit saw, some flour (which they say Terapuhi, the Grey chief, had sent to him by Mr. Mackley of Nelson), and some potatoes from the Maori garden at the mouth of the Teramakau.

Jan 14: Prospecting up the Ohono Creek about four miles, got a few colours in several places, cutting a small race in the reef near our camp at the mouth of the creek to get some water for sluicing. The four men who were with us before arrived this afternoon. We gave them a little flour, and advised them to make up the beach line [the route along the beach] for the Buller or return to Christchurch, as there were no provisions down here. They left us next day to take along the beach line.

Jan 15, 16: Prospecting in the creek and cutting a race and making a saw-pit.

---

1 Blue marl: a mudstone containing a mixture of clay and calcium carbonate, formed under freshwater conditions.

2 'Country' is used here to mean district.

3 The Little Hōhonu (or frequently simply the Hōhonu). Smart used various spellings.

4 Epiphytes: plants such as mosses, lichens and ferns that take root and grow on a host tree.

5 'To go mates with': to go into partnership with one or more diggers, sharing the labour and any gold retrieved. Sealed with a handshake, it appears to have been regarded as a verbal contract and legally binding. See Smart's account of Nicholas Saint Dizier v. James Anderson, page 83.

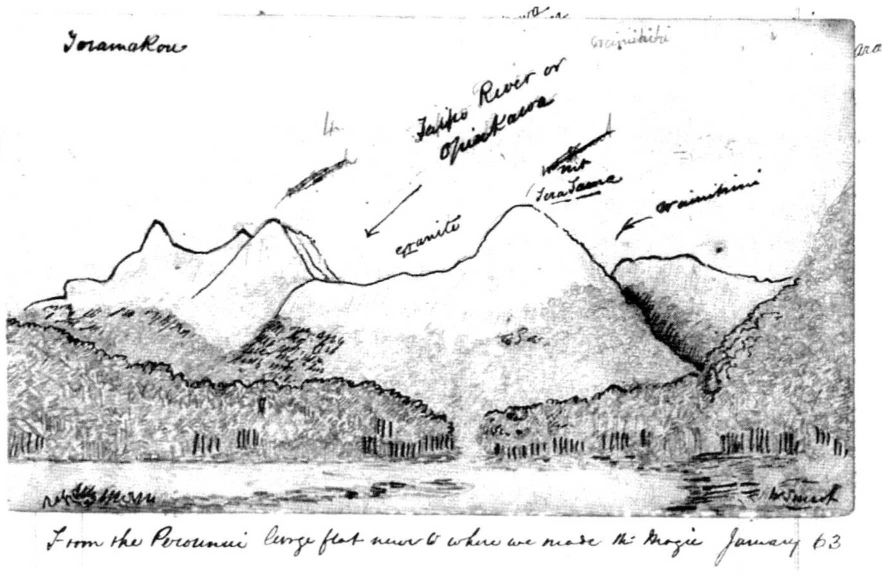


***Teramakou [Taramakau]. W. Smart***
*From the Perounui – large flat near to where we made the Mogie, January 63.*

Smart has indicated the 'Taipo River or Opiakoura [Hopeakoa]', and 'Mt Tera Tame' [Tara Tame], on the south side of the Taramakau. At top right he has written faintly 'Wainihihi' [Wainihinihi, a tributary of the Taramakau], and below this he has drawn an arrow pointing to the position of the Waimakariri River, east of the Main Divide. He has noted 'granite' in the centre of the sketch.

Jan 17: The Maories returned this afternoon with a decoy-bird, a kaka, and a Maori boy, Hione by name, a saw, and potatoes from the garden. In a small canoe, Jack [Ihaia] Tainui and the boy went across the river to some old totara trees and caught 58 kakas with snares, and some fish.

Jan 18: Sunday. North-west wind, hot and fine. Saw a shoal of fish but could not catch them with a hook. The sandflies are very bad, in myriads.

Jan 19: Put up a saw-pit for cutting boards for sluice boxes. Davis, Money, French and Everest all came down on a raft,[1] having followed on our tracks, and very short of provisions.

Jan 20: Made a sluice dam, cut some planks and made some boxes for sluicing, with nails we had brought with us. Maories working in our party.

Jan 21: Working at the mouth of the Ohono. Everest and two of the Maories went to the Grey to see if they could buy some flour from Terapuhi out of Freeth's stores, as our stock was nearly done. We had saved it as much as possible for travelling

1 See Charles Money, *Knocking about in New Zealand*, Capper Press, Christchurch, 1871/1972, pp. 50–55, for a marvellous description of this voyage down the Taramakau.

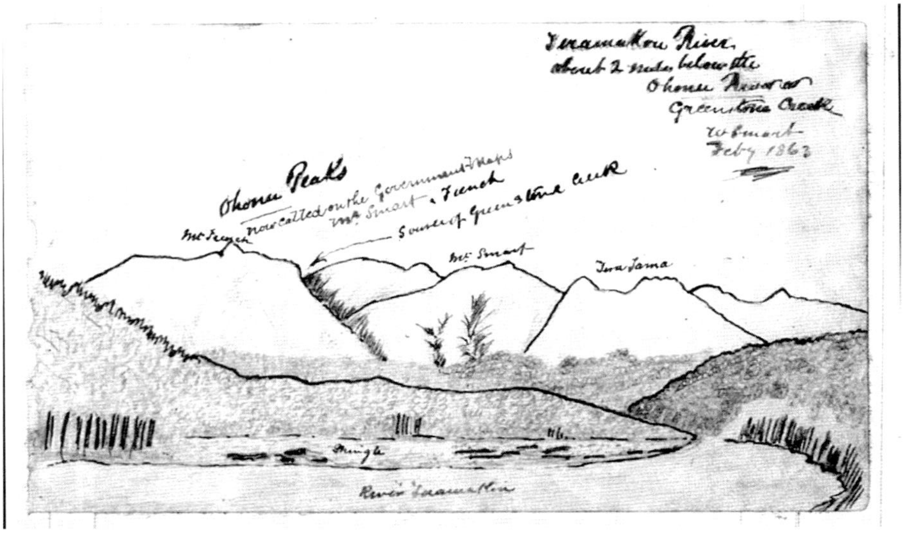


***Teramakou River, about 2 miles below the Ohonu River or Greenstone Creek, W. Smart, Feb'y 1863***

Smart has identified the 'Ohonu Peaks' [Hōhonu Range] and 'Tera Tama' on the right. The names of the peaks 'now called on the Government Maps Mt Smart and French' and the 'Source of Greenstone Creek' were added later (1890?). In the foreground Smart noted 'shingle' on the far bank of the 'River Teramakou'.

purposes, as potatoes would be too bulky and heavy to do a journey on.

Jan 22: Sluicing wash from the bank of the Ohono Creek.

Jan 23: Sluicing. Everest and the Maories returned with a hundredweight of flour and some potatoes, and he had a job to get the old chief to sell it.

Jan 24: Sent letters and some gold by a Maori who came up from the Grey to go to Kaiapoi.

Jan 25: Sluicing. Gold very fine and scaly.

We have come to the conclusion that it will be useless to stay here any longer, as the Maories have no quantity of potatoes, and unless some means are adopted to get a vessel round with stores before winter sets in we shall stand a chance of being half starved. We also want to go further up the creek to get on to better gold, and with the few tools we have we cannot work satisfactorily.

The sandflies are a terrible torment.

Jan 26: Patterson and Peter started early this morning and went a long distance up the Ohono and up the right-hand branch. It was dark when they returned.

They say they have found a large greenstone boulder. This boulder was, some

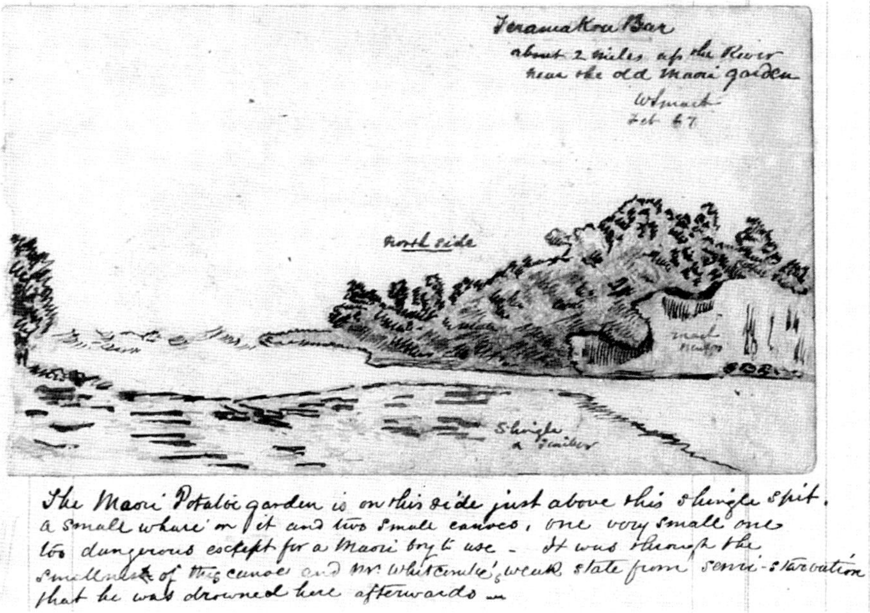


***Teramakou Bar about 2 miles up the River near the old Maori garden, W. Smart, Feb 67***
*The Maori Potatoe garden is on this side just above this shingle spit, a small whare on it and two small canoes, one very small one – too dangerous except for a Maori boy to use. It was through the smallness of this canoe and Mr. Whitcombe's weak state from semi-starvation that he was drowned here afterwards.*

Smart has indicated the 'North side' of the river, and 'Marl Bluff', centre right. On the right (foreground) he noted 'shingle & timber'.

years afterwards, the subject of a law suit in the Supreme Court at Hokitika. It was attempted to be taken away by some white men; the Maories claimed it and sued them for £1500. They gained the case and got the stone back.[1] The place it was found was the celebrated Maori Point on Greenstone Creek, as this was the name given to [the upper section of] the Ohono afterwards.[2]

1 It appears that Peter and Patterson did not claim ownership of the boulder, perhaps because they did not belong to the Poutini Ngāi Tahu hapū. It was a local man, Simon Tuakau, and others who later claimed customary rights to it. The Pākehā James Reynolds, who came upon the boulder, regarded it as 'finders keepers'. When Tuakau took possession of it Reynolds took him to court. After two trials and two hung juries in Hokitika, the case went before the Court of Appeal in Wellington, which Tuakau eventually won. *Southland Times*, 3 August 1866, 2; *West Coast Times*, 8 August 1866, 2; 9 August 1866, 2; 10 August 1866, 2; 11 August 1866, 5; *Daily Southern Cross*, 31 October 1866, 6; 9 November 1866, 6.

2 As part of the Ngāi Tahu Treaty Settlement of 1997, the Hōhonu River/Greenstone Creek was renamed the Greenstone River/Hokonui; a tributary retains the name Little Hōhonu River. The Big Hōhonu, a separate river flowing from the Hōhonu Range into Lake Brunner, has since been renamed the Eastern Hōhonu River.

Jan 27: The Maories believe we shall have heavy rain and a big flood, so we today shifted our camp from the low creek bed on to the high banks of the creek. Everest's party did not shift their camp as they were advised. It came on to rain in torrents about 9pm.

Jan 28: Everest's party was washed out, and had to come over to our camp in the night. At daylight their tents were under water to the ridge pole. On the rain ceasing, the flood soon went down again.

Jan 29: Our flour being all gone, we all agreed to go down to the Grey and see the old chief about stores [in order] to do a little more prospecting before leaving, as we were confident that if we had tools and provisions we could get some good gold in this creek.

We started down the river about seven miles to the beach on our large mogie, which had broken adrift, but found it washed up about a mile lower down the river. Went to the Maori garden, cooked a good feed of potatoes and then crossed to the beach.

An old Maori pah on the south side of the Teramakau and a few miserable huts on the north side; their occupants a few very old women and some children half eaten with sandflies, who were somewhat fearful.

We too were nearly eaten alive with fleas and sandflies. These sandflies are a most dreadful torment: to pan a dish of dirt your hands are covered with blood. We have to tie our trousers at the ankles, and to sit still for one minute is impossible. One comfort is they are away at dark, and then the mosquito begins, but he is not so bad as the sandfly.

From here we walked along the beach line to the Grey River, about ten miles, crossing the Paroa[1] at low water and some few smaller creeks. The beach is covered with small quartz pebbles, black sand and small red garnets, and minute specks of gold in places – 'flour gold' – as fine as the finest snuff, and will float on water when dried in the sun.

All bush down to the high-water mark. We crossed a creek running from a beach lagoon into the Grey, and following a track through the bush and scrub. Reached the Grey pah, which is a clearing on the south bank of the Grey of, I suppose, about four acres in extent. There are five or six huts built of timber and grass and toitoi, and a number of stands for drying eels and other fish, some futtahs[2] for storing provisions, about one acre of ground in wheat stubble, and some patches of potatoes, taro and maize. Several canoes, tools, nets, etc.

---

1 The Paroa River, also known as the New River, has been renamed New River/Kaimata. The crossing on the beachline would have been about 2 kilometres north of the Taramakau river mouth.

2 Smart's spelling of the Māori word 'whata', an elevated platform mainly used for storing food.

Old Terapuhi, a very hospitable old Maori and the Grey chief, gave us everything he had. He made us welcome, gave us the use of a nearly new whare to stay in, and plenty of eels and potatoes.

We had a long talk round a large fire outside at night, burning some coal which the Maoris had brought down from a few miles up the river where they had been for eels and wood-hens. Splendid coals for steam purposes could be got to the beach at no very great expense either by tramway or punts.

The old man says he is the elder brother of Werita Tainui of Kaiapoi, and that when he was a boy there was a large pah on the north beach of the Grey and a great many natives lived there, but they died off and some went further south. The old pah is nearly all washed away by the sea encroaching on it, and they then shifted to the south side of the river. He has several sons-in-law and daughters; some of these were away at the Buller.

He described the different routes overland to the East Coast: one by the Teramakau, one by the Otira and another by the Hokitika and down the Rakaia. By this route, he said, many years ago about 20 Maories died in the snow.[1]

The North Islanders [Māori from northern tribes] used to come down the beach line in numbers, killing all [local] Maories and searching for greenstone, and the natives used to go up the rivers to hide from them when there were too few to resist.

The old man said he had been over all the country from Nelson to Lake Wanaka. He had no knowledge of gold but said he has seen similar stuff to what we shewed him in several creeks and rivers, and would go back with us to the Ohono to see how it's found,[2] and he would sell us as much flour as he can spare.

1863, Jan 30: Bought some flour. Money and Davis started by the beach line for the Buller. I went across the Grey River in a canoe to look at the remains of the old pah with Peter and Patterson.

Peter thought it was 26 years ago since he was here last. Some totara posts and fencing was still sound. Here the last old chief,[3] a short time before

1 Browning Pass, an old greenstone route on the Main Divide between the Arahura River in the west and the Wilberforce, a tributary of the Rakaia, on the eastern side of the Alps; it was called Noti Raureka by some Māori.

2 Tarapuhi may have been deceiving Smart; as Ihaia Tainui's uncle, it seems unlikely he was ignorant of gold digging.

3 This was Tuhuru Kokare, chief of Ngāti Waewae, a hapū or subtribe of Ngāi Tahu. He founded the Poutini Ngāi Tahu hapū based at Māwhera, and was father of Tarapuhi (who succeeded him) and Wereta Tainui, the Kaiapoi chief. Maika Mason, 'Tuhuru Kokare', *Dictionary of New Zealand Biography, Te Ara – the Encyclopedia of New Zealand*, www.teara.govt.nz.

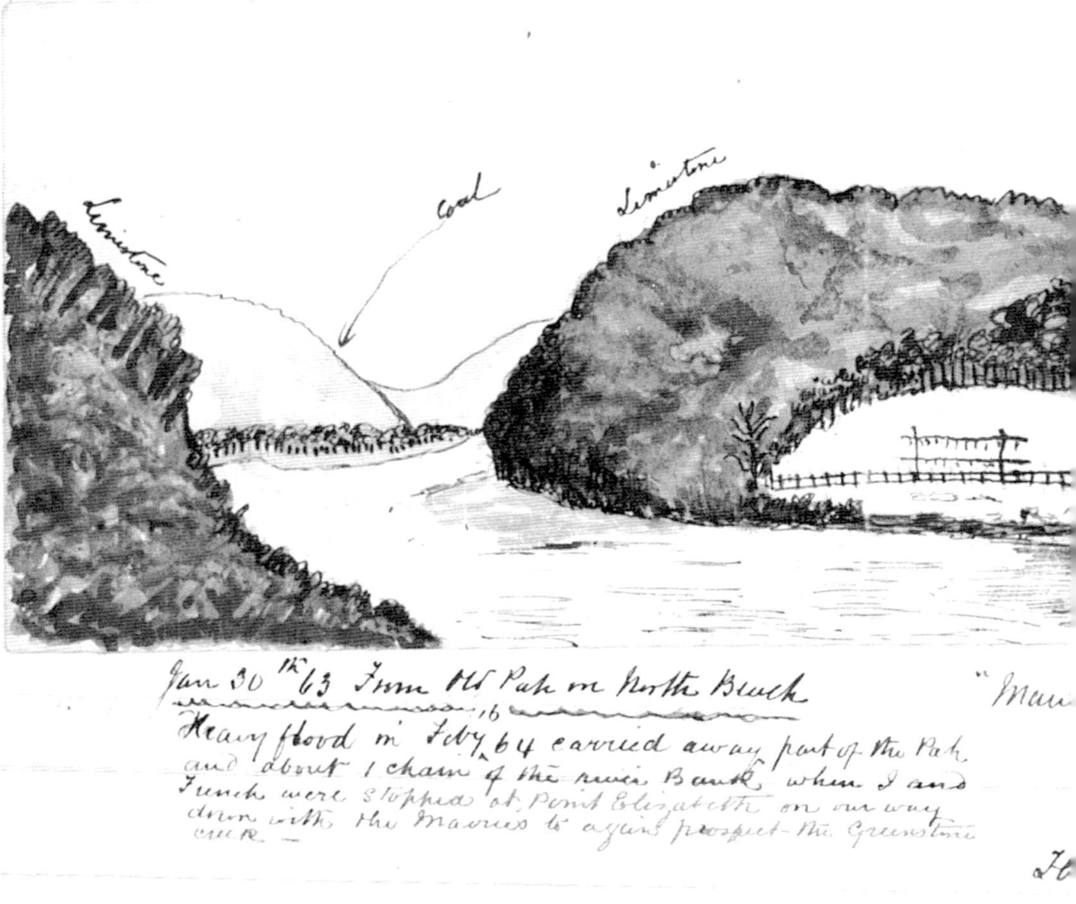


***Mouth of River Grey, Terapuhi's Pah, Mawhera, south side, Jan 30 '63***
*From Old Pah on North Beach. Heavy flood in Feby 64 carried away part of the Pah and about one chain [20 metres] of the river Bank, when I and French were stopped [by the same storm] at Point Elizabeth on our way down with the Maories to again prospect the Greenstone Creek. Canterbury Govt Built Store in May 63 under Townsend. Jan 64 W. Revell sent over in charge of store. Feb 64, J. Revell, French, myself and the Maories removed the store to the lagoon, W. Revell being away to ChCh.*

his death, buried his greenstone mere [hand weapon]. The Maories have searched for it many years without success.

The Grey has a very good bar and deep water in the river, close to huts in the pah on the south side.

Old Terapuhi remembered seeing me in Kaiapoi about four years ago when

he was over. There were no other white men but ourselves south of the Buller,[1] nor stores, except what he had belonging to Freeth, and he believed he, Freeth, would come overland in a few months with sheep.

Jan 31: Bought some flour. Terapuhi and the rest of us left for the Ohono, reached the Teramakau and crossed in small canoes to the potato garden where we camped for the night.

Feb 1: We walked up the river, the Maories taking our swags in the canoes. Met a survey party from Christchurch: Mr. Drake, Richard Sherrin[2] and Solomon the Maori [Horomona Tuapaki] as guide. They had come down the river to

1 This was so far as he knew at that time; there were actually a few others.

2 Jakob Lauper was also in the party. En route to the Coast, Drake had met him at Taylor's station, where he had been recovering from his recent ordeal, and hired him to join the expedition.

see about a track, but were returning again to join Howitt's party who were again cutting the track to go round by Lake Brunner.

Feb 2: Prospecting, and the Maories up the creek looking for greenstone. Shewed Terapuhi how to wash gold with a dish.

Feb 3: Prospecting, sluicing, got a few pennyweight of gold. But with no hope of a vessel arriving and provisions getting short, we thought we had better make a start for the Buller and get a vessel to come down or wait until the Canterbury Government sends round stores for the survey party, which they must do if the Coast is to be surveyed at all. All the white men had left for the Buller some days since, with the exception of ourselves and three others who agreed to go with us.

Feb 5: Returned to the Grey pah with Terapuhi. He advised us to go on the overland track – up the Grey River, across a saddle and down to the Buller,[1] and prospect on the way. He agreed to go with us as far as possible with two canoes, which we hired from him. Rain at night.

Feb 6, 7, 8: Rain. At the Grey.

Feb 9: Our party now consisting of seven. We started up the Grey River with Terapuhi and two canoes, carrying plenty of eels, potatoes, salt and very little flour. After paddling and poling about seven miles, we came to the coal seam, cropping out in the banks of the Grey River, a large one of splendid quality. But it would have to be brought down in flat-bottomed boats, or a very costly road made to it; in fact an enormous sum of money [would be needed] to open out coal in such outlandish country.

We camped on the south bank of the river and it began to rain. On leaving the Grey pah, we passed some limestone rocks. Old Terapuhi pointed out a place in them where there is a cave in which all their old chiefs were buried – but no one had been buried there for years and the place was overgrown with saplings and scrub. The river trending well to the north. Rain at night.

1863, Feb 10: Rain. River in flood. Caught four wood-hens and some eels. Had to shift our camp on account of the wind.

Feb 11, 12: Camped. Rain, and river high.

Feb 13: Started this morning, but too much water in the river to go far, so again camped.

Feb 14: An early start. Crossed the river to the north side and left the two canoes with Terapuhi to take back to the Grey pah. Through bush and riverbeds all

1 This inland route would take them up the Grey and Little Grey rivers to the Inangahua and on to its confluence with the Buller River.

day and small creeks. Camped about 6pm.

Feb 15: Crossed the river to avoid a high terrace, via a good ford, and on to some open grassed country above the junction of the Ahaura, about 20 miles up the river from Greymouth. This is the country taken up originally by Lee. Freeth put sheep on the country afterwards, but it was too wet and most of these died.

Followed up the river on this side through several open patches of country; made a very long day's walk. Crossed the Grey four times and reached the country taken up by Mackley at the Little Grey, where we camped near the main stream. Can get a fine view of the Grey Valley from here.

[Our camp was] on a flat at the head of the Grey in the forks – one branch running to the Waiau Saddle, the other to the Inangahua Saddle, a tributary of the Buller River. The Grey runs through a wide bush valley in a serpentine course over rough shingle bed with only one gorge about five miles from the beach. Two large tributaries coming in from the southwards, viz the Arnold from Lake Brunner and the Ahaura. Three large creeks on the north. The open country consists of about 8000 acres in patches divided by belts of bush, finely grassed, about 20 miles from the beach, also about 5000 acres at the forks at the head of the Grey.

***Head of the Grey Valley. W. Smart 1863 Feb 15***
*This country taken up by Mackley as a Sheep Run.*

Smart has labelled, left to right: 'Pass to Waiau Gorge', 'Broken Bush[?] Hill', 'Source of Grey', 'open country', and faintly, at top right, 'Valley of the Grey'. In the centre, at the foot of the hill, he has noted, 'Terrace, open country', and in the foreground, 'Grass Flat'.

Feb 16: On up the Little Grey through scrub, bush and river, and prospecting. Camped on the river bank at night.

Feb 17: On up the riverbed which was getting narrow and rough. A great scarcity of birds here and we were getting very short of provisions: one pannikin of flour a man per diem. Held a consultation and we thought it best to return to the Grey pah and take up the beach line. If we go on we have a saddle to cross and many miles down the Inangahua to travel before we reach the Buller, and if the birds are as scarce over there as here, we shall be starved. So we returned to our previous night's camp. Sandstone reefs, granite and slate wash and in all probability gold bearing.

Feb 18: Followed the river down to the first flax swamp and all hands started to make a mogie.

Feb 19: Finished the mogie and came down the river. Found it difficult to steer down the falls, and to avoid the snags sticking out from the banks. We had to be very careful and get underneath them. In coming down one fall we saw a very long snag sticking out from the bank which we could not avoid; going under this, we had to jump over, and let the mogie go underneath. Our front men cleared it nicely but it swept myself and Fraser backwards into the river. But we were both alongside the mogie again at the bottom of the fall, and being a splendid warm day our clothes soon dried.

Shortly afterwards another snag capsized Fergusson and a lot of our swags. These we recovered again, with the exception of a billy. However, we managed to arrive at about dark at the place where we camped on the first night coming up from the Grey [pah], and with very much less labour than walking.

Feb 20: After going a few miles down the river this morning we met some Maories coming up in a canoe. These were Terapuhi and some others and one of his sons-in-law, McLean,[1] who had just come down the beach line from the Buller, and reported that the Maories were getting some good gold at Lyell Creek, and they were on their way back again via the Inangahua Saddle. Terapuhi had left his brother, Werita Tainui, in charge of the Grey pah. He had arrived overland from Kaiapoi.

They wished us to return again with them [to the Buller], but we preferred going by the beach line. So we left our mogie and returned with Tainui in the canoe to the Grey pah early in the day. We made a good stock of pararas [pāraerae] (flax shoes which we had used for some time, our boots being

1 Nothing more is known of McLean, except that he was mentioned as being with the group of Māori miners who took part in the early Greenstone rush of mid-1864. *West Coast Times*, 9 August 1866, 3.

quite worn out), and got everything ready for an early start in the morning.

Feb 21: Crossed the Grey in canoes and started out. Crossed over the bluff at Point Elizabeth and camped.

We had to put ourselves on short rations, one small scone per day for each man, which was baked every night and served up to each man next morning – this he could either eat at once or make it last the day. For tea we used bidi bidi [piripiri], a small creeper covered with burrs, which we had been using for some long time as our tea was all done. Rather a bitter taste, and unfortunately it created an appetite. We had salt and knew we could get plenty of mussels off the rocks if we got no birds, and in places we could eat the top of the nikau palm which grew near the beach line.

1863, Feb 22: Very fair travelling today along a great stretch of beach; in places great quantities of black sand and garnets. All bush and scrub down to high water. Camped near a bluff, a bad place for fresh water; had to walk some distance to get it.

Feb 23: Crossed the bluff and thence over the Miko Cliffs, travelling nearly all day in the bush. We kept inland instead of going down the flax ladders on the cliffs, as the Maoris told us the old ladders were rotten. Camped on the beach at night.

Feb 24: Very rough travelling all day, over rocks and boulders principally. Camped in a large cave.

Feb 25: Over rocks and boulders all day. We have evidently missed our track and ought to have turned off inland yesterday. The scenery along the beach line is very grand; some beautiful little bays, but we were too hungry to properly appreciate it.

Feb 26: Over rocks and boulders, the route getting worse.

Feb 27: Crossing rocks until we could go no farther, we then had to turn inland by getting up the cliffs by the aid of flax and scrub. We got onto the top of the cliffs, and after heading inland for about a mile we came to the Maori track on an open piece of level country where we ought to have been two days previous. We followed this track, and then down a river and on to the beach line again and camped.

Feb 28: Crossed round a point and along a sandy beach when we were stopped by a river with a heavy flood in it, although we had no rain on the beach line; it must have been raining in the back ranges. We here caught several young Paradise ducks – flappers – by running them down.

Mar 1: Made a raft with some dry logs and some flax, and crossed the river, then along a sandy beach until stopped by another river. Waited here for low tide, and then we forded it near the mouth. We then travelled on until we got to a point and some rocks. Here we camped, waiting for low water. Got up in the middle of the night and got round these and again camped until daylight. These rocks I thought could be the Steeples from their appearance; I therefore knew we could be no great distance from the Buller River.

1863 Mar 2: We rounded another small point and then lit a fire and ate up all the balance of our flour. We then started again and after about an hour's walk could see the mouth of the Buller River and a store on the beach; and to our surprise, when we got to the banks of the river, we found a party of Maories there with a canoe waiting for us. They had seen our fire, which we fortunately happened to light in a place where the Maories generally light a signal fire when coming up from the south.

On reaching the other side we were heartily welcomed by both whites and Maories, and Reuben Waite the storekeeper cooked us a good dinner of the best he had, of which we stood very much in need as we had had very scanty fare for the last 6 weeks.

Mar 3, 4, 5: Taking a spell and a look round the place. The stores here are three in number and kept by Waite, Isaac Blake and John Martin; wooden buildings of no great size. The Maories own the land and the storekeepers pay them a rental. They are on the north side of the river and the Maori pah is just at the back of the stores on open ground just clear of the bush.[1]

The beach line here is rather wide and consists of large scrubby flats and terraces, the range of hills lying back from the beaches both south and north some three or four miles.

On the south side there appears to be a large extent of open ground with grass in patches and rushes. Waite talks about taking this up as a cattle run, but I suspect [the animals] would soon get away wild into the bush. This flat was afterwards found to contain leads of gold running through it and was called Addison's Flat.

There are a good many Maories here and some fine-looking men amongst them. Tamati Freeman[2] is their chief, one of the Maories who went with Te Rauparaha to attack Kaiapoi years ago; also old George at the Lyell. Most

1 These few buildings were the first in the settlement that became Westport.

2 Tamati Pirimona Marino, aka Tomati or Tommy Freeman, was chief of one of the Aorere pā. His tribal affiliations were Ngāti Mutunga, Te Atiawa and Ngāti Rarua. Gottfried Lindauer, who painted his portrait, wrote that Marino had joined Te Rauparaha's raid, and added: 'as his share of the conquest he received a large area of land at Massacre Bay [Golden Bay]'. Hilary and John Mitchell, *Te Tau Ihu o te Waka: A history of Māori of Nelson and Marlborough, Vol. IV: Ngā whānau rangatira o Ngāti Tama me Te Ātiawa: The chiefly families of Ngāti Tama and Te Ātiawa*, Huia Publishers, Wellington; Wakatū Incorporation, Nelson, 2014, pp. 411–21.

***Mouth of the Buller River 1868, W. Smart***
*The Maories had a large Hut and Garden on the South side of the river but they lived on the north side.*

Smart has labelled the 'Maori house & garden' on the far bank of the river.

of them are away at the new-found diggings at Lyell Creek. I believe there are only about 30 whites altogether, who are at the Lyell and several other parts of the Buller river and at Waimangaroa. These men come down in their canoes for stores, both whites and Maories, and take a stock with them to last some time.

No goldfield is proclaimed here. Neither are there any police or Nelson government officials, except Mr. Skeet, a surveyor, who is cutting a track up the south side of the Buller; he has a few men at work for him and draws his stores from the storekeepers here month about, giving each one an equal share [of his business].

Mar 5: Terapuhi, the Grey chief, with the rest of the natives we left going up the Grey on February 20th, arrived here in a canoe. They had had fine weather, and the old chief, who had many years ago left a canoe in the bush on the banks of the Inangahua River, had found it again. Although it was overgrown with moss and scrub, he had used it to come down the river in.

Doctor Mackley and Fred Moore had come down with them. They had just reached Mackley's run with sheep, which they had driven over by the Waiau River [over Lewis Pass], when they met the Maories and had come over with them to take up some stores via the Buller and Inangahua to the run. Moore had engaged with Mackley to help him with the sheep at the station.

Mar 6 & 7: Myself and Day went to the Waimangaroa Creek about nine miles north of the Buller. There were only two or three men at work here as all the others had gone up the Buller to the Lyell. It is a very rough creek – slate country, large boulders – and some good nuggety gold has been got here. The creek rises in the Mount Rochfort range where the men say there is plenty of coal.

Returned to the Buller and found some Maori canoes had come down from the Lyell Creek. Tommy Freeman shewed us some splendid nuggets got by the Maories from there; every party of Maories it seems has to pay him a certain tribute of the gold they get.

Mar 8 to 21: Very bad weather; the river in constant flood so that no canoes can go up to the Lyell. Went up to the Canterbury garden, as it was called, in the canoe made by Jackson's party when they were down here in the schooner *Fantome*. Martin the storekeeper went with us.

On the 21st the schooner *Gipsy* anchored off the bar waiting for a wind to come in. She and the *Mary* are the vessels which bring stores down here from Nelson. Heavy fresh in the river.

Mar 22: Heavy N-west wind and sea. The schooner *Gipsy* got under weigh and tried to beat off the coast, but after some hours she was obliged to run on shore on the sandy beach near the Steeples. The passengers, who were miners, walked along the beach to the Buller, but there was such a heavy sea running in the river that the Maories stripped and took their largest canoes to bring them over.

Mr. Skeet's foolhardiness however was the cause of two men drowning as, contrary to all persuasion, he would go out in his canoe, taking two men with him to pull [row] and Charlie, the half-caste, to steer. He crossed the river and took in three of the passengers and when he was about two-thirds of the way back again, the canoe capsized. One of the passengers, John Chadwick, and one of Mr. Skeet's men, Taranaki Jack (Jordan), were drowned. Although both of them were splendid swimmers, we saw them go down like a stone and were powerless to aid them.

The rest hung on to the bottom of the canoe and were taken off her by the Maories just before she got to the bar. Chadwick had 100 sovereigns on him but neither of the bodies were ever found although the beach was searched for miles.

Most of these men went up to the Lyell; among them was Pat Sweeny and Docherty.

We again saw Money and Davis who had been out at Waimangaroa. They had not lost their peculiarities in being lightly clad, each carrying very light swags. Davis' eccentricity lay principally in boots which were seldom fellows [a matching pair]. At the same time I believe he could tell you the history and doings of nearly every English politician for a century back.

Several men came up from the Grey River in ones and twos, nearly half starved. The Maories and storekeepers sent men down the beach line several times to see if they could pick up any travellers who were coming over and set up some markers they could travel by.

1863, March 23, 24, 25, 26: Too much water in the river for the canoes to go up.

Mar 27, 28, 29, 30: Left this morning for Lyell Creek, with three canoes and Maories. The river is very rapid and without a ford from its source to its mouth, sometimes running between high precipitous walls of rock and deep water; at other places opening out into large flats but all covered with bush to the water's edge; in other places swift-running falls.

The canoes were poled and paddled and towed by ropes, running up in the backwater eddies, and then crossing the river to another eddy; then pulled up a fall by all hands, except the steersman, having to stand up to their knees in water at the top of the fall.

It is a very tedious and dangerous journey, and requires great care to prevent capsizing; and if the canoe takes a sheer,[1] the current is so strong the men cannot hold her and over she rolls. Several white men have already lost their lives, and one or two Maories. We passed one or two diggers' camps on the river; one was at the long fall. The road party's camp was not a great way up, and Simpson had charge of the stores there.

There are 38 falls between the Buller mouth and the Lyell Creek, which is about 45 miles by the river and about 18 miles in a direct line from the beach; but the country is so mountainous and full of deep gorges and canyons that a track could not be made in the direct line.

Mar 31: We arrived at Lyell Creek this evening and it is a very bad crossing, I think one of the worst on the river – it is just between two very bad falls. Camped at the mouth of the creek.

April 1: Went up Lyell Creek with August and Charley, prospecting. Took up a portion of the creek, fixed our tents, and started to put in a wing dam when

1 Takes a sheer: swerves, or changes course suddenly.

***Buller River above the Rock Fall. W. Smart, 1863***
*This is a nasty fall [rapid] to go down in a Canoe. There is another one called the sluice box fall lower down which is very rapid. It sometimes takes 10 days to come up this river from the mouth to Lyell Creek, and generally about 6 hours to go down with a good Maori crew.*

Day, who had agreed to start a store at the mouth of the Lyell, wanted me to come [in] with him instead. He gave [me?] my share of the claim,[1] and went down and built a store with the assistance of some men. This was the first and only store there for some considerable time. On more miners coming, and the splendid nuggets which they were getting there, other storekeepers were induced to come up: John Felix (Maori Store) came, Blake's Store, kept by John Lang and Knyvett, and also Simpson from Skeet's road party.

The Lyell Creek is a deep gorge creek many miles in length; in many places the sun never shines in its bed, granite and slate rocks, and it is fed from the Lyell ranges.

All hands were working the creek bed by turning the half of the stream on

1 Smart's exact meaning is unclear. Smart soon joined Day in running the store.

***Lyell Creek, Smart & Day's Store & Camp, 1863***
*This was the first Store built on this Creek. A large tree came down and smashed part of this to pieces, so we had to build another on the opposite side of the Creek, years after all the points were worked for gold.*

one side while they worked the other side. The Maories were getting some fine nuggets and they were mostly brought to me to weigh. I had some to weigh of all sizes up to 54oz.

Some of these Maories were very superstitious. I knew one, Big Benny, who was once in the Native Police in Wellington. He would say, 'Our party are going to get a big nugget tomorrow or the next day, I know. I had a certain dream last night, and we always get a big nugget when I have that dream', and sure enough he would be down with a nugget of up to 25oz.

There is a tributary of the creek called Irishman's. It had been prospected by several but nothing got out of it until a party of six Irishmen set into [it] and worked it all up on a face.[1] These men got there 1100oz in about four months and they brought one nugget to show me, the biggest; it weighed

1 To work up on a face: to expose a vertical working surface (a cliff, terrace or tunnel) for mining by excavation or sluicing.

108oz – the shape of a small dumb-bell.[1] One of the men, Red Jack, afterwards opened Red Jack's Creek[2] on the Grey River.

Arthur, the coloured man who was with us in the Teramakau, came up here and as he was very hard up I gave him leave to stop with us in the store and go out prospecting occasionally. After many weeks without getting anything, he came home one evening and said, 'Well Boss, I got the first good colour today, let's have a drink! You see, Boss, I traced the fornication[3] of de gold right up to de big Hill.' He then pulled out of his pocket a big nugget wrapped in a bit of calico. It weighed nearly 52oz.

I lent him enough money to take him down the Buller and on to Nelson, where he sold the nugget and returned some weeks later with earrings in his ears and rings on his fingers, after buying tools and everything necessary to work again. He paid me all he owed, then went back to work in the same ground, but found nothing to speak of afterwards near where he got this nugget. This was up in the hills towards the big Manuka Flat.

[Smart's use of daily diary entries stops here, resumes in early February 1864, and is intermittent after that.]

We found some black sand and gold at the junction of the Lyell Creek and the Buller, and I made a quicksilver cradle,[4] and various kinds of riffles[5] to work it. But we did no good with it as the gold was so very fine, similar to the beach gold at the Grey River. Jack Bartle[?] and his mates went further up the river and got similar ground, and saved some gold, but it was difficult to save.

Most of the miners here were fond of exploring and prospecting, and some

---

1 At over 3 kg, this would have been the largest nugget ever found in New Zealand. Smart may or may not have weighed it himself. Numerous very large nuggets were found in the Lyell district around this time. Rumours of the discovery of one weighing 108 ounces made headlines in the Nelson newspapers during October and November 1863, but its actual existence was never verified. Accounts of gold finds were often exaggerated or inflated by rumour. At the end of November 1863 the Bank of New Zealand in Nelson purchased a large Lyell Creek nugget shaped like a dumb-bell, weighing 52 ounces. *Colonist*, 24 November 1863, 3.

2 To open: to discover a promising piece of ground and publicise it to encourage storekeepers and miners to come and make it a successful goldfield.

3 Perhaps he meant 'formation': Smart wrote 'fornication', but whether it was a mistake or a joke on Arthur's part is not clear.

4 Quicksilver cradle: a wooden box placed in a sloping position and fixed on rockers, in which gold-bearing gravel is washed; the gold is caught by mercury in the lower part of the cradle. A sluice-box, on the other hand, is a long sloping trough with grooves on the bottom, into which water is directed to separate gold from gravel or sand.

5 Riffles: strips of wood or other material used to trap gold in a sluice-box.

used to go away for several weeks at a time to the higher branches of the Buller, and on the ranges when the weather was favourable.

I and some others were on the banks of the Buller above the Lyell one day, where we came across 4 men who had come overland from Nelson. They were nearly starved. They had eaten their dog, and must have shortly died, had we not met them.

Several men were drowned during the time I was here, from being too venturesome; I believe quite 20 lost their lives in this manner. There were three sailors who made a flat-bottomed punt at the mouth of the river, which they loaded up with stores. After a very long passage up the river they capsized at the Lyell crossing and two were drowned; the other one reached a rock in the river and was taken off by a canoe. Simpson was also drowned at this crossing.

The miners here were Stevens, Cameron, Healy, McGaw, George Cundy, Nicholson, Forrester, Davis, Stone, Dalgetty, French, Carter, the Watsons, Big Bell, Ferguson, Ryan, Harris and some other white men, and several Maories. We made our own rules as to the size of claims etc, and sent one man to Nelson for robbing his mate of a nugget, for which he got six months' imprisonment.

We had a vigilante committee of 12 men to make certain regulations, as to size of claims, and to settle mining disputes. I acted as chairman. We tried two or three bad cases, and it would have gone hard with some of the culprits had they not been restrained by some of the quieter men.

One man was sent away in a canoe with orders never to return, under pain of being lynched. Another, Scottie [James Anderson], I had great difficulty in saving from having the rope around his neck. This Scottie robbed his mate, a German [Nicholas Saint Dizier], of a nugget weighing over 50oz. He had planted it in the creek near our store, but he was watched and seen to unearth it the evening before the provision canoes left. In the morning we stopped the canoes, tried him there and then, and sent him to Nelson in the charge of two men. He got six months, and all his costs were paid out of his share of the nugget.[1]

Mr. Blackett of the Nelson Engineers Department and Dr. Tatton paid us a visit, and Dr. Husband was working here and lived with me for some months. He was afterwards with the well-known Captain White of the *Blue Jacket*. Some years after, I met Dr. Husband in Christchurch, an enormously stout man; when there [at the Lyell] he was as thin as a lath, and it took me some time to credit he was the same man.

1 See *Nelson Examiner & New Zealand Chronicle*, 24 September 1863, 2.

Several men made their way here towards the end of '63, amongst them F. Moore and W. F. Warner of the Commercial [later Warner's] Hotel in Christchurch. I persuaded him and two others to go on to the Manuka Flat and sink a shaft. They got down about 30 feet, but met with such large boulders that they had to give it up. So after that they cleared out overland and went on to Nelson and then back to Christchurch. Warner shortly after took [over] the Golden Age [Hotel], after being with Ruddenklau at the City Hotel. Warner, myself and others tried to work the black sand at the junction of the Lyell and the Buller, but did not make it pay and a lot of it was carried away in a big fresh.

Another coloured man came, Johnson, who had been cook at the Royal Hotel in Christchurch. He was with Howitt's party cutting the track, and from him I heard about Howitt's tragic drowning in Lake Brunner. Howitt was camped alongside Lake Brunner; he went out one night to catch eels, leaving Johnson alone in the camp. Howitt never returned, and Johnson stayed hunting about the lake, until he got short of provisions; he then came on past the Grey on to the Buller and to the Lyell in November '63.[1]

We were completely isolated with regard to news from anywhere, as we never have a letter or paper. A Maori brought me one day a brass plate which had an inscription in memory of Claude M. Ollivier and asked me to forward it to be placed on his grave at Jackson's Bay, which I did, but who sent it I never heard.

The Buller River was sometimes very heavily flooded, and on one occasion it rose 80 feet above the ordinary level in the gorges, and backed the water up the Lyell Creek over the top of the counter in Johnny Felix's store. Of course at these times canoes could not travel and we were consequently hard up for food – we had very little indeed.

Once, myself and E. Knyvett lived for nearly a fortnight on split peas, salt, a little tea and sugar only. Some men had only a few potatoes per day and there was no means of getting relief until the canoes came up. There was a big rush for stores when they did come; the frying pans and billies were going for some time to satisfy the men's hunger.

The Maories brought up the stores at £45 per ton, which made everything very dear. We never had fresh meat, and numbers of the men got ill with

---

1 Johnson had briefly visited Howitt's camp at Lake Brunner in May 1863, en route to the diggings. He was ill-equipped, starving and cadging food. It was not he, but James Hammett, a member of Howitt's work party, who remained behind when Howitt went out eeling on the lake with two others of his work party. When all three failed to return, Hammett searched for days for his lost colleagues. Later, Johnson met him on his way to the Buller and relayed Hammett's tragic news to the people at Lyell Creek. Charlton Howitt, 'Journal of H. C. Howitt', MS-1038, Alexander Turnbull Library; *Nelson Examiner & New Zealand Chronicle*, 29 August 1863, 5.

scurvy; some of them had to come down to the store as they were unable to cook, and I fed them with potatoes and turnips and plenty of vinegar, which soon cured them. I had quite a hospital at the store and many men afterwards told me that I was the means of saving their lives.

Yankee Dan carried up stores also, in a canoe named the *Goldseeker*. There were very few birds about this creek, but the rats were very numerous and it was a trouble to keep any provisions out of their reach.

The rats were so numerous and cunning that they got at everything they could. They would even try to take a lighted candle stuck in a bottle. I used to put a candle in a bottle on the floor and lie in my bunk and shoot them with a revolver.

Our old friend Reuben Waite brought up about a ton of Wanganui bacon. It was so bad that no one would buy it so he hung it all up in an old hut. When he next came in, it was all gone except the bones and the skin – the rats had bored holes through it all over; a few pieces at all eatable were taken by the diggers, but the rats had the rest. Davis made a comic song on it, to the tune of *Wait for the Wagon*: 'Waite brought us bacon and we all took a side'.

The Nelson Government served all the storekeepers here a nice little trick to raise some money. They sent a man down called Harper. He worked in the creek and about the stores for nearly four months and was presumed to be a digger, until in January '64, Mr. Mackay and a police sergeant arrived at the Buller from Nelson in the steamer *Sturt*.[1] They came up the river to the Lyell in a canoe and arrested this man, representing that he was wanted for something in Nelson.

They went down the river a few miles so as to get him out of the way. The policeman then returned the following morning and served all the storekeepers with a summons to appear at the Buller [Westport] on 6th February '64 for selling spirits without a licence. Other storekeepers at the Buller were served the same.

A court was held before Mr. Mackay, and Harper was brought out as the informer. All the storekeepers were fined £50 each; but for the store I represented, the fine was not enforced as I had fortunately sent through Waite for a bush licence several months previous. Wm Everest who had a billiard table at the Buller was also fined. (This man and his mate were afterwards smothered in their tent by a landslip from the Caledonian Terrace.) By this raid the Nelson Government cleared about £300. They also

1 It was a detective who accompanied the police sergeant to the Lyell, not James Mackay Jr, the Resident Magistrate at Collingwood; but it was Mackay who later presided over the court proceedings at Westport. *Colonist*, 9 February 1864, 3.

took with them a North Island Maori who was down trying to enlist recruits for the Hauhau fanatics, and Fraser for selling a gun to a Maori.

[January 1864] Terapuhi, the Grey chief, came up to the Buller and asked me to return with him to the Grey, as the Canterbury Government had a store there to supply the survey party, and the men who were cutting the track, and that I could get stores and go again to the Greenstone Creek.

Simon [Himona Tuakau], Ihaia Tainui and some other Maories had been working occasionally at our old workings at the mouth of the Greenstone, where we had worked 12 months previously, and they had built a hut there that I could occupy. They had also found some coarser gold in working around the greenstone boulder, which was originally found by Patterson and Peter. He [Simon] intended to take back [from the Buller] some drills and a hammer to try and break up this large greenstone boulder as he considered that of as much value as the gold.

1864, Feb 9: I left the Buller for the Greenstone by the beach line, with Terapuhi and four Maories who had been to the Buller on a visit, and Michael French. The Maories carried with them some drills and hammers, and several articles which could not be got at the Grey. We had nice weather for five days and found several short-cuts across points which we had missed in coming up the coast 12 months previous.

It was much easier travelling, except our ascent of the Miko Cliffs on flax ladders, which was far from pleasant; the flax ladders had to be renewed. Our swags and a dog they had with them were pulled up by ropes.

On nearing the Grey on the 15$^{th}$, it came on to rain heavily so we camped on the beach near a lagoon just before we reached Point Elizabeth. One of the Maories went on without his swag to let the Grey natives know we were coming. We were camped under a cliff and just beside us was the lagoon with piles of drift timber all about. We had large fires in front of our tents and had all turned in comfortably, but it was raining very heavy.

About 11pm I was awoken by feeling water under me. I woke the Maories and French who were sound asleep and by the time we got our clothes on we were up to our knees in water. We took some burning sticks from the fire before they were extinguished, and our blankets, and made our way to a large pile of logs in the lagoon which we got upon and all sat there together covered with our wet blankets until daylight. It was raining and blowing heavy from the N. West and we thought that daylight would never come.

When the day broke, the wind and rain ceased and we crossed to another part of the beach, made some huge fires, dried all our clothes, and rescued our tents which although they had been under water were tied fast to some pegs. One of the Maories went to the Grey River and back again, telling us that we must stop there for the day as the river was so very much in flood that a canoe could not cross, and he could see that a portion of the pah was washed away.

1864, Feb 17: Arrived at the Grey and found the river still very high. Tainui crossed us in a canoe and we found about a chain of the river bank had been washed away, also eel stands, a whata, and the whare we had camped in previous; also the water was undermining the government store [the Grey depot], which had been built since we were there.

Mr. J. [John] Revell was in charge of the store, his brother, W. H. [William] Revell, having gone overland to Christchurch on business. We learnt that this store and its contents were sent round from Christchurch in May '63 under the charge of Mr. Townsend. Captain Buxton in *Wild Wave* [*Crest of the Wave*] brought them round after calling in at Wellington for some salt beef. They came on to the Grey but could not find the entrance to the river, as they had got too far south and were off the Hokitika River. Captain Buxton, Townsend and some others left the vessel in a boat to find the entrance, and after pulling some long time, found they could not return to the vessel, so they beached the boat on the south side of the Hokitika River. They dragged the boat into the river and crossed and camped in a Maori hut on the north side. Next day, after properly securing the boat, they walked up the beach to the Grey. The schooner arrived, landed the cargo and they put up the store.

Some time afterwards, Townsend, Peter Mitchelmore, William Sherrin and Solomon the Maori[1] went down to the Hokitika to bring up the boat, and in crossing the bar at the Grey, the boat was capsized and Townsend, Peter and Solomon were drowned. Sherrin managed to swim ashore. The body of Solomon was not found.

After this was known in Christchurch, Mr. W. H. Revell[2] was sent over to take charge of the store in place of Mr. Townsend in January '64, in the *Mary* from Nelson.

Mr. Whitcombe, surveyor, had also been drowned. He had left Christchurch in April 1863 and had found a pass on the Rakaia Saddle and come down the main branch of the Hokitika River, he and his only companion,

1 Plus Simon (Himona Tuakau) and two other Māori.

2 Before his appointment as manager and storekeeper at the Grey depot, Irishman William Horton Revell (1829–1893) had been a member of the armed constabulary in Kaiapoi. After the start of the West Coast gold rushes, he became the government agent, goldfields warden, resident magistrate, coroner and returning officer. He also marked out Hokitika's first two streets, one of which is named after him. Mark Pickering, *The Colours*, 2011, p. 158; Julia Bradshaw (ed.), *The Diggers' Story: Accounts of the West Coast gold rushes*, Canterbury University Press, Christchurch, 2014, p. 39.

Jacob Louper [Jakob Lauper]. They were nearly starved on the trip, and Mr. Whitcombe was unfortunately drowned in attempting to cross the Teramakau River in the small canoes which were there. Jacob was carried out to sea and came back with the tide. He managed to get to the potato garden, and from there followed up the Teramakau to our old camp at the mouth of the Greenstone, where fortunately for him some of the Maories were camped. They put him on Howitt's track, where he met the men who were there at work cutting the track, and from there he went on to Christchurch by the Teramakau Saddle.

[In February,] Mr. Revell asked us if we would help him shift the store, as in the event of another flood the store would be carried into the river; this of course we agreed to do.

Mr. John Rochfort and Mr. Arthur Dobson were expected to finish the survey of the Coast, and Mr. James Drake and Richard Sherrin were finishing Howitt's

***Part of the Miko Cliffs, about 40 miles North of the Grey River, West Coast, NZ. Wm Smart 1863***
*The Maori track followed up this Gully where marked, then along top of the Cliff some distance and down a flax ladder on to the Beach again. The Nelson Govt afterwards had a chain ladder put up on this track. The two large rocks outside [in the sea on the left], which were once a portion of the Cliffs, are timbered on top also. The trees grow quite down to high water mark, and plenty of fresh water all along the beach line from the small creeks.*

Smart has noted: 'Sandstone cliffs', 'creek', 'Maori track' and his 'tent'.

track. As soon as this work was finished, they had received instructions to leave the Coast, as no one here had as yet found payable gold; and none of the men cutting the track had found payable gold. In fact there has never been any gold found on the track cut by Howitt's party. They did not follow down the Teramakau as we did, but went off inland round by Lake Brunner.

Further prospecting was not to be encouraged. Men who came through were to be allowed to purchase only sufficient stores to carry them on to

Christchurch or to the Buller; however, as we were going to help shift the store, he [Revell] would let us have sufficient stores to do some prospecting. If it had not been for myself and French, this country would have been abandoned, perhaps for years.

We chose and cleared a place to put up the store, quite away from the river on the banks of the lagoon on a low terrace. The Maories floated up the timber framing and the stores in empty casks and canoes, and we fixed it all up again.

While this work was going on, Mr. W. Revell and John Smith returned from Christchurch on 29th March '64 by Howitt's track, the route of which, after coming over the Teramakau Saddle and down the Teramakau some distance, turned to the right to Lake Brunner; thence across into the left-hand branch of the Greenstone Creek, then down the creek, crossing it about eight times, to its junction with the Teramakau; thence down the north bank of the Teramakau to the beach.

After finishing the store, we, with Messrs Revell, chose a place for a cemetery, not far south of the Grey on the beach line terrace, where we buried the remains of Mr. Townsend and Peter Mitchelmore, which we got from their burial place north of the Grey River, and the remains of Mr. Whitcombe, which we brought from the Teramakau.

After this, French and I got some stores and went prospecting up the Paroa and on the beach line, but the gold was very fine. We then went back to the store and got as many provisions as we could carry and went up to the Greenstone and camped in the hut built on the site of our old camp by Simon the Maori.

While we were here, Dobson's men came down in April '64 overland from Christchurch with two pack horses and a mule to do the survey.[1] The men from Drake's party also came down; they had finished Howitt's track and were going to the Grey to get some stores to carry them on to Christchurch where they had to go to get paid. They said they would return and do some prospecting with me up the Greenstone, but they had found nothing on the road they had cut. One of these, quite a young man, was Albert Hunt.

French and I then went back up the Greenstone, leaving some flour at different places on the road, hung up in trees. We had very bad weather; the creek was constantly in flood. We got gold with the dish in numbers of places, good payable prospects for sluicing, but patchy. After being up the creek for 22 days we returned to Simon's hut, where next day arrived Hunt

1 This was Arthur Dobson, who had returned to complete his survey contract for the government, begun the previous year.

and his mates from Christchurch. They started to work in the bank of the Greenstone Creek, our old workings, and put through a good-sized paddock using our old boxes for sluicing. They got about seven oz for the eight days' work; however, Hunt's mates would not stay any longer and returned to Christchurch, and Hunt went down with us to the Grey store where we all camped.

We found we could not depend on getting supplies from the Government Store, as they were sometimes ordered to sell and next month *not* to sell.

During our absence prospecting, Mr. W. Revell had again been to Christchurch and back, and had again received instructions to abandon the Coast. I told him that we had found gold which I thought payable, and that I and French were going to the Buller to get a vessel to come down and bring stores and miners back with us.[1]

[On 24 May] I wrote a letter to Bealey, the Superintendent of Canterbury,[2] informing the Canterbury Government that I had found payable gold and claiming the reward. I gave him [Revell] this letter but it was delayed in being forwarded and I got no reply for more than 12 months. I received very bad treatment with regard to this gold discovery as I am certain the Coast would have been abandoned if I had not sent to Nelson for stores.

Mr. Arthur Dobson was surveying [to the south] and Mr. Rochfort surveying at the Grey, laying out the Maori reserve, and Terapuhi had been helping to cut the survey lines.

Before leaving for the Buller, Terapuhi asked me to go up to the Waimatuku Creek for a few days' prospecting and he took French and me and a Maori across [the Grey] in a canoe. On our return we were to fire a gun and a canoe would be sent to us.

We were away several days and on our return found that old Terapuhi was dead.[3] Messengers were sent to the Buller, to Kaiapoi and all over the place to collect all the natives together for the funeral. So we of course delayed our return to the Buller, and camped in our tents at the pah. A number of Maories soon came over from Kaiapoi and other places on the Coast. Tommy Freeman [Tamati Pirimona Marino] and seven men came down from the Buller; some of them brought kegs of sherry wine.

On their arrival, a track was cut from the pah to the burial cave in the

1 Smart and French took these deliberate steps to trigger a rush to Greenstone Creek, and thereby prove it to be a payable goldfield.

2 In fact he addressed the letter to William Revell.

3 Tarapuhi died on 8 April 1864.

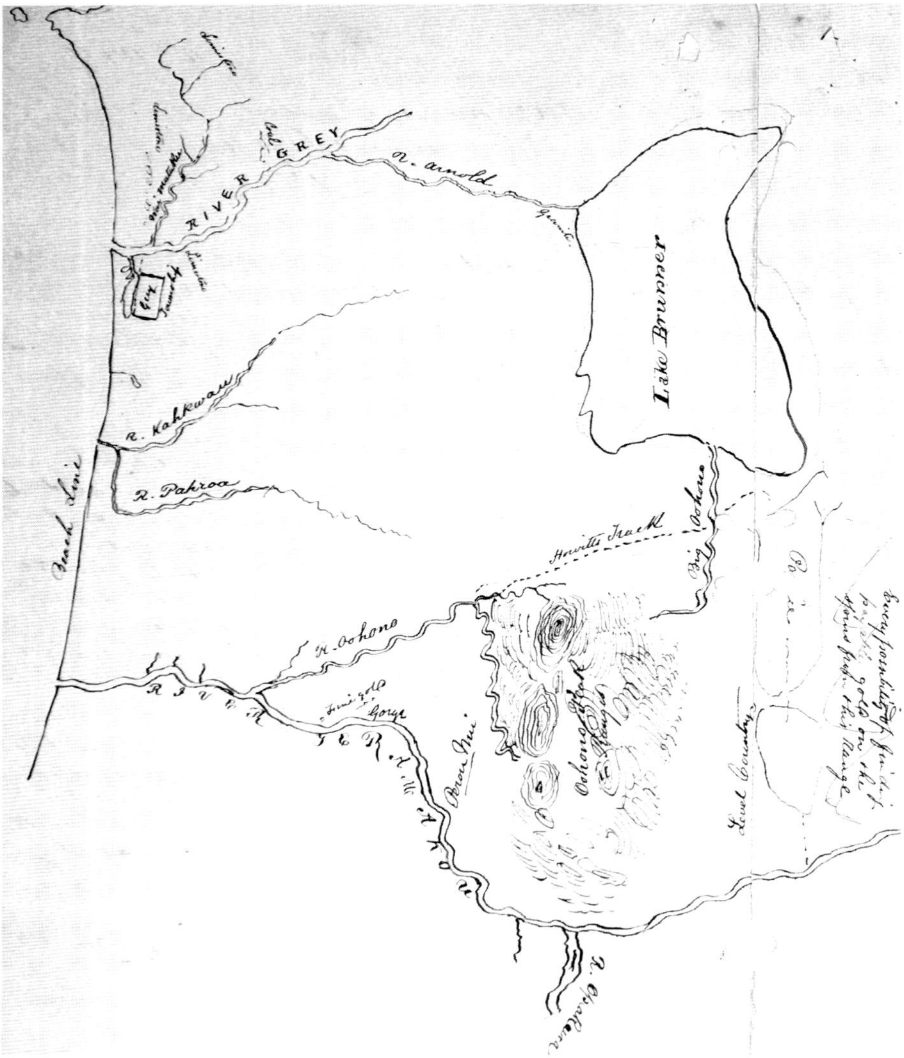

Smart and French attached this 'rough sketch of Country where we have been prospecting since February' to their important letter to William Revell. It shows the 'Grey township', site of the government depot and Māwhera pā and, on the coast, the 'beach line'. Beside the Taramakau River, Smart has noted 'Tainui gold' near the 'Gorge'. He indicated the positions of the 'River Oohono' (Hōhonu), with Greenstone Creek (unnamed) running at right angles to it; the 'Perou Nui' river flats; and the 'Level Country' between Lake Brunner and the Taramakau, an area known as the 'Pakihi' (obscured; 'pākihi' means grasslands or open country). Howitt's track can be seen leading from Lake Brunner to the 'Oohono', crossing the 'Big Oohono' river. The 'Oohono Peak/Range' dominates this landscape, and Smart has noted: 'Every possibility of finding payable gold on the spurs from this Range.' The dark circles show the approximate positions of Mt Smart and Mt French (not yet named) above the Greenstone Creek where the pair found payable gold. See also map on page 123.

William Smart and Michael French to William Revell, 24 May 1864, in Revell, W. H., 'Correspondence relative to claims made by Albert Hunt and others for the reward offered for the discovery of a payable gold field in the province of Canterbury', 11 August 1865, Canterbury Provincial Council, Papers and Returns, CAAR CH287 20410 Box CP 618 PPC 9, Archives New Zealand, Christchurch.

limestone rocks. A ladder was made and two Maories went down to prepare a place for the body, and very frightened they looked as they came up again.

The funeral service was read in Maori by George Peters,[1] and the body lowered into the cave with all the deceased's clothes, sticks, and everything that he used, even to a pannikin, and very glad the Maories seemed when they were once more in daylight.

After the funeral, the Maories had a feast to which all the people there were invited and which was carried out with great ceremony. The white men who were there were Messrs Revell (2), Rochfort, French, John Smith, Williams, Mr. Rochfort's man, Albert Hunt and myself. Mr. Dobson and his men being away, these were all the whites on this part of the Coast.

I told Tommy Freeman about the gold and he instructed two of his Maories to stop behind and go up the Greenstone, whilst the rest of us went on to the Buller. We got all the stores that Mr. Revell would allow us to buy and gave them to Albert Hunt to go up the Greenstone with the Maories, prospecting. Hunt behaved very badly to us and also the Maories afterwards. Revell treated us very unfairly in his report about the gold discovery.

1864 June 1: Left for the Buller to send for stores. We left the Grey pah at 1pm: Tamati Freeman, myself, French and 15 Maories, Messrs Revell and everyone else coming to the river to see us off.

We made a very quick trip, as we had little to carry as the Maories had left some stores on the track for their return journey. We walked day and night to suit the tides; and on June 4th in the evening we got to the Buller, a distance of about 90 miles. We got some splendid crayfish at the Miko Cliffs, off which we had a remarkable good feed with potatoes cooked in a copper Maori.[2]

June 5: Had a long talk with [storekeeper] Isaac Blake, who agreed to go to Nelson with the first boat that came in and charter the schooner *Mary* to take us and about 30 men and stores down to the Grey. During his absence we agreed to get a building all framed ready to take down for a store. A craft came in about a fortnight after and Blake started in her for Nelson. I also sent a letter to Reuben Waite, who was in Nelson, and also a sample of gold from the Greenstone Creek.

---

1 This was George Peter Mutu, the son of Peter Mutu. In 1872 he was ordained at Kaiapoi as a deacon of the Anglican Church. Henry Jacobs, 'Colonial church histories: New Zealand', Society for Promoting Christian Knowledge. http://anglicanhistory.org/nz/jacobs_histories1887/04.08.html. Accessed 18 March 2016.

2 Copper Maori: a colonial term for a hāngī or umu (a traditional earth oven). The origins are unclear, but it is possibly a corruption of 'kapa Maori', as distinct from a 'kapa Pakeha', an iron pot like an English copper and used as a boiler. See H. W. Orsman (ed.), *Dictionary of New Zealand English*, Oxford University Press, Auckland, 1997, p. 169.

This letter of mine was published in the Nelson papers[1] and copied to the *Lyttelton Times* as follows:

> **Gold Digging at the Grey**
>
> We have been favoured by Mr. Waite with the perusal of a letter from Mr. William Smart; it contains the following:
>
> 'I have found very fair prospects at the Grey[2] but could get no Provisions from the Government (Canterbury) Store as their Agent has orders not to sell any more.
>
> I got in several places four grains to the dish. It was gold similar to the old diggings (Buller) gold; and I consider it far before the Buller, for we could get gold wherever we tried. I saw a 'paddock' put down in the banks of the river for trial; it was in very indifferent ground (part of our old ground [at] Ohono Creek), and the men knew where much better places were.
>
> They were eight days at work in the paddock and got seven ounces. Three of the men left for Canterbury and one is now stopping on the claim living on potatoes till a craft comes with provisions.
>
> The country is very accessible and plenty of feed for pack bullocks. The Maories have horses down there and so has Mr. Dobson and his party and these are used for packing on their survey.'
>
> So satisfied is Mr. Waite with the above information, that he is now in treaty for chartering the steamer *Nelson* for a trip to the Grey for the purpose of taking with him a large quantity of provisions.

This copy of my letter is a trifle altered from the original: mine was 'three grains to the dish'. The men knowing better ground, feed for pack bullocks, and the Maories having horses are little additions of my friend Reuben. The rest is correct.

While waiting for Blake's return I sent to the Lyell Creek and got 30 men ready, and the store framed, and Day and I went to Waimangaroa Creek and got Watson's party also. French and some others wouldn't wait for the vessel and went down [to the Grey] overland.

We were all getting tired of waiting, expecting to see the schooner *Mary* every day, when to our surprise on July 20th 1864 in came to the Buller the steamer *Nelson* with Waite and a crowd of men. We then heard for the first time about the Wakamarina diggings [in Marlborough]. Nelson being full of men, Waite soon had a number to come on to the Grey so had chartered the

---

1 *Nelson Examiner & New Zealand Chronicle*, 12 July 1864, 5.

2 'The Grey' was a widely used umbrella term that included the Greenstone diggings.

*Nelson* for that purpose, and he had called in at the Buller to pick up some more stores. Two or three canoes were put on board the steamer and she cleared out for the Grey the same evening and got into the Grey on July 22nd 1864, being the first steamer to cross the Grey bar.

We were much wanted [by some of the passengers] to go down in the steamer, but as we had arranged with Blake we were bound to wait for the *Mary*, which came in a few days afterwards. We had a splendid passage down to the Grey and anchored off the bar, where we were stuck outside for six days at anchor without any wind to go in.

The next day the steamer *Nelson* on her second trip [to the Grey] towed us in. This delay lost us some ground which we wanted up the Greenstone, and which was all taken up by Maories and others and was afterwards called 'Maori Point', where the greenstone boulder was found by Peter and Patterson in '63. At the Grey, Waite fixed up his store at the Maori pah, and Blake built his on the south beach between the lagoon and the sea.

Some of the men who came in the first steamers were regular roughs and, because they could not get on the gold at once, were going to kill or hang everybody who had caused the rush. Waite's store was rushed and Blake's nearly battered in with boulders.[1] They turned Hunt off the Greenstone Creek on account of a letter he had written which was printed in the Otago newspapers.[2]

Hunt persuaded some men to jump [seize] some of the Maories' ground at Maori Point, himself armed with a gun to assist. The Maories, not wanting to have a general row, gave these men a portion of their ground. This was deep ground in the creek, and it turning out badly they fell out with Hunt also.

These Maories had treated Hunt very kindly and took him up with them when I and French left for the Buller in June to get stores brought down. One of these Maories, George Peters, is the Maori parson at Kaiapoi; his father, Peter Mutu, and Patterson, were the finders of the greenstone at Maori Point when they were with me in January '63.

Also during this time, French and myself were prospecting in the One Mile Creek south of the Grey. This was all very fine gold, and our tent was pitched on a terrace point where the two small branches joined. In this point the Watsons got afterwards the 105lb weight of fine gold.

1 See page 118 for an alternative view of these events.

2 The letter was first printed in the *Lyttelton Times* of 6 August 1864, and then syndicated to other newspapers including the *Otago Daily Times*. In it, Hunt boasted that 'a party of four might make £150 a week' at the Greenstone – an outlandish sum.

[July 1864] Mr. W. Revell had again gone to Christchurch and taken some gold which had been got by the Maories and Hunt. Constables Broham and Cooper came back with him in August, and the Canterbury Government then learnt that a gold-field was being gradually opened out. But they were not at all pleased to see it, according to the *Lyttelton Times* of 30th July '64: 'as a gold field is being thrust upon [the people of Canterbury] we may console ourselves that it has turned up in the remotest corner of the Province'.[1] The Christchurch people did all they could to stop the working of this field for fully 12 months.

A number of the roughs and dissatisfied men who had gone away and written to the papers stating that there was no gold in the country were very glad to return when they heard that people were getting very good claims in the Greenstone Creek. This was the first creek [in West Canterbury that miners had] started to work, and men are still working there now (1890).

The first store to start on the Greenstone was Ferguson and Buchanan's, in the hut built by Simon, which was added on to. Another store at the junction of the left-hand branch was kept by the Macfarlane brothers.

1864 September: Myself, Day and McGaw went over on the south side of the Teramakau and made a canoe to go up the river prospecting. Healy, Stevens and Cameron's party were going also, Healy having been up the Taipo River and got a good prospect of coarse gold.

We finished the canoe on September the 9th and loaded her up with stores and went up to Ferguson's store at the Greenstone, where we camped several days. It was raining and a good deal of water in the river. Some of our mates canoed up several loads of stores for Ferguson, which were packed along the beach from the Grey on horses. We then went on up the Teramakau with Cameron's party in three canoes.

We took our canoes into the mouth of the Taipo River and dragged them up into the bush in some thick scrub. We then pitched a spare tent we had in the bush about 100 yards back from the river and hung up in it all our spare stores. Cameron's party took their small canoe up the Taipo River with stores, through the gorge to the mouth of the Seven Mile Creek. We worked here some weeks and Cameron's party flumed[2] a piece of the creek and worked the bed up on a face. All the gold was coarse reef gold, every piece being attached to quartz.

---

1 The point being that those who disapproved of a goldfield within Canterbury could be content that its evils would be far away and have little effect on their lives.

2 Flume: to divert, by a flume or water-race, the waters of a stream in order to lay bare auriferous ground.

While we were there, the men at the Greenstone heard we were away prospecting and a regular rush occurred. They followed us up the Teramakau and camped at the mouth of the Taipo. It coming on to rain heavily and the Taipo being in flood, they came no further, and getting hard up for provisions they crossed the Teramakau on to Howitt's track and returned to the Greenstone Creek.

Hunt and Donnelly came and worked for a few weeks in a creek above the Seven Mile. They fortunately did not find our tent nor our canoe, although they were camped very close to it. Most of them were not used to bush country and seldom went far from the banks of a river or creek.

After leaving the Taipo we went up the Teramakau and up the Otira as far as the gorge, but we there found nothing. We then returned and crossed to Lake Brunner and prospected round these granite ranges, but got nothing. While there we met Reed and Stone who had come over the Teramakau Saddle by Howitt's track with a few head of cattle, which they were taking on to the Greenstone; these were the first cattle brought overland.

On November 29th we left by Howitt's track for Christchurch, where we arrived on 7 Dec '64. I received in Christchurch a letter in reply to the letter which I wrote in May '64, claiming the reward for myself and French, to the effect that [the goldfield] was not accessible to Christchurch, signed E. Jollie, Provincial Secretary.

The only recognition we ever received for our efforts in opening out this goldfield was the naming of two mountains at the head of the Greenstone Creek: Mount Smart and Mount French.

The *Handbook of New Zealand Mines* by [William] Larnach, Minister of Mines, published in 1887, gives a very inaccurate account of our doings in connection with the opening of this gold-field, also about the greenstone boulder and other subjects.[1]

By the end of 1864, numbers of places were opened out and Hokitika was growing into a township. The Maories and old hands had always some men out prospecting.

In January 1865, Mr. Robinson, Superintendent of Nelson, Fowler, McCulloch and Cook were drowned in trying to land from the steamer *Wallaby* at the Buller bar. He [Robinson] had come down to see about the coal-fields at the Buller and the Grey.

On 1st March '65, the West Coast of Canterbury was proclaimed a goldfield by S. Bealey, Superintendent.

1 For more of Smart's disparaging comments about the *Handbook*, see box on page 108.

Public meetings were held in Christchurch and Kaiapoi to urge on the Government to make a road to the Coast; [the Government being] still in doubt about it being a payable field. Different routes were tried. Mr. Browning went over the Rakaia route where the Maories had been lost in the snow, and which I described to him as well as I could from Tainui's information. With him were Griffiths, Money and Newberry. Mr. Browning put a paragraph in the paper [stating] that he had received information from me re the route he went, now called Browning Pass.[1]

Mr. George Dobson[2] went to open a track by the Otira in March '65, which route was afterwards decided on after his report dated 28th May '65.[3] Mr. Harman also did some exploring by the Rakaia. On 16th March '65 the Government published a caution to diggers not to travel overland as a great many had been drowned and nearly starved.

In April '65 Dr. Haast was sent over the Hurunui route to report on the goldfield; see report, 4th May 1865, in the *Lyttelton Times*. Mr. Sale was appointed Commissioner for the Goldfield and Mr. W. Revell [Goldfields] Warden.

Rushes were taking place all over the Coast. Banks were established and their agents buying gold. Mr. Walmsley[4] had come down here from the Buller; he was buying gold there in '63. I got several parcels for him from the Lyell Creek at that time.

On the 17th July '65, myself, Day and Fraser again went to the Coast by the Hurunui Saddle. The rivers were very low and no snow on the Saddle. We went round by Howitt's track into the Greenstone and then went on to the Grey and stopped in Blake's store. On the 1st of August we arrived in Hokitika and camped until the 18th. We had heavy squalls, and during this time several vessels of all sizes came ashore [ran aground] on the beach.

August 19th: Went up the Hokitika River to the Kanieri where there were a good number of miners and stores; from there to Woodstock, and up the Kokatahi

1 This pass was considered for a time to be the best route for the new road across the Alps. Work started on the road in 1865 but was soon abandoned.

2 Brother of Arthur, and also a surveyor.

3 Arthur Dobson, on the advice of Tarapuhi, had found the Otira pass in 1864. In 1865 the government scrambled to find a less challenging route for the new road. After reviewing all known options, they chose 'Arthur's pass'.

4 Walmsley was the Bank of New Zealand agent on the Otago goldfields. He was transferred to Nelson, and sent to set up a branch of the bank at the new settlement of Hokitika at the beginning of 1865. Philip Ross May, *The West Coast Gold Rushes*, Pegasus, Christchurch, 1962, p. 104; *Cyclopedia of New Zealand, Wellington Provincial District*, The Cyclopedia Company, Wellington, 1897. http://nzetc.victoria.ac.nz/tm/scholarly/tei-Cyc01Cycl-t1-body-d3-d15-d6.html.

to near the saddle,[1] then back again to Hokitika on September 4th, where I met Mr. Browning, Money and Newberry who had just come over by Browning's Pass.

On September 7th we went on to the Grey, where there were rushes going on every week to the different creeks. We camped on the 8th all day at the Teramakau as it was raining heavily, and on the 9th arrived at the Grey and camped in the bush. Met Watson and went down to their claim where they got 105lb weight of gold near my old camp (where I camped in February '64).

Vessels were frequently coming ashore at the Grey; the steamer *Nelson* went on to the north beach on the 23rd September '65.

Day and I went to work on the beach line lower down, nearer to Paroa. The beach leads were just beginning to be worked along the Coast, both north and south of the Grey; some were very narrow, only a few feet wide.

We were on this and the north beach until December '65. We went to try the large quantities of black sand we previously saw at the Seventeen Mile Beach, but it was excessively fine: a large copper cradle plate when covered all over [with it] weighed a mere trifle.

Met Mr. Walmsley who had been stuck up and robbed near No Town[2] by armed men; he lost about 800oz of gold and £1000 in notes. The packer, who was with him with about the same sum on his horse, escaped. He, Walmsley, was robbed once before at the Waimea, his valise with about £1000 in it being stolen from the tent he stopped in.

In December, the Maories told us they thought they had got on some good gold on the north side of the Grey above the Ahaura; so we went up with about 40 Maories and they took us across in one of the canoes to an island where we camped. Before daylight next morning 3 or 4 of them went to look at the creek. They stopped all day and returned at night as they found it was not good enough. A large rush of men followed the canoes up, but we all came down the river the next day.

A great many of the diggers went on up the Arnold track to Lake Brunner, when there was a big rush on to the Kangaroo Creek, but we went with some of the Maories to the Waimea.

Large sums of money were made during these rushes by men who had made bridges or had a boat on the different creeks or rivers. Dan Macfarlane, one of the storekeepers at the Greenstone, had a ferry on the Grey which was

---

1 Browning Pass, which was reached from the Kokatahi via the Styx River. Another route to the pass was by the Arahura River.

2 On Twelve Mile Creek, a tributary of the Grey, not far from Stillwater.

afterwards known as Langdon's. He took on one rush nearly £1000 for a week's work.

The man at the bridge across the entrance to the Greymouth lagoon also made a lot of money; men were constantly rushing backwards and forwards – very few kept steady to work at one piece of ground for the first two or three years.

The beach rushes were in full swing at this time, and some very rich patches were got all along the coast – some in the higher terraces and some down to the high-water mark. By December '65 these were already worked out.[1] The Darkie's Terrace near Point Elizabeth was rich, but they, the prospectors, were cheated out of part of their ground by Dufty tapering them off with 16ft cut out of the middle of the tape line. One of the Darkie's mates was Godfrey, afterwards living at Brunnerton.

Watson's was the first rich patch, and some lower down to Paroa, then the Auckland Lead and on down to Hokitika; then the rich ground at Okarito.[2] But the best of this ground did not take more than six to 12 months to all work out, except where the leads ran into the swamps and deep ground, and no doubt large quantities were lost and left in the tailings.

The Watsons came down with men on the *Mary* from the Buller. I laid them on to the creek about a mile south below the Grey, where I had got some fine gold with French. They worked up the creek a short time and made about £1 a day, then they left to go to the Greenstone. They again returned from the Greenstone to the same creek, and one day in cutting into the bank came across a seam of black sand about four feet thick which glistened with gold. They saved this in tubs and bags about their camp until they could get some quicksilver [mercury] to amalgamate it. They afterwards carried the amalgam to Hokitika. Mr. Pardoe retorted it for them and it weighed 105lbs weight. This was the first start of the black sand leads. These men lost a lot of gold in their tailings.

[In 1865] the Canterbury Provincial Government decided on making the road by the Otira Gorge, this being the most central route from the plains; but the saddle [Arthur's Pass] is much higher ground than that by the Hurunui and Teramakau. [The work] was let in sections under different contractors.

The contractors in the Otira were supplied by stores which were principally packed over the Hurunui Saddle on Howitt's track. William Alexander had a

1 Worked out: mined until depleted of gold or other minerals.

2 Smart, like many others at the time, spelt this 'Okarita'.

store at the junction of the Otira and Teramakau. He packed [in provisions] and supplied some of the men there. He afterwards started a store about 6 miles lower down the Teramakau, now (1890) known as Jackson's.

By this Otira route, mails were sent over in July 1865, partly by coach and pack horses, and the Maories then canoed them from the Taipo to the beach.

The Provincial Government, after in the first instance doing nothing to assist in opening a goldfield, went to the other extreme and [in December 1865] actually started an escort under Inspector James to bring gold over to Christchurch. They built barracks and stables on the road at intervals, and spent a lot of money most foolishly; a more absurd venture was never thought of. The escort made only one trip, and some of the diggers offered them a few pennyweights to take over in ridicule; however, the buildings came in useful afterwards for swaggers and the coach proprietors.

The Government made a deal of money out of the duties on gold and customs during 1865. The more men that came to the place the better, as the country required numbers to prospect it and the gold was all over the Coast.

1866 January: We were in Greymouth on the north beach, and in February we went up the Greenstone Creek. We there saw French, who was in a store at Maori Point. Lots of claims had been worked out, both in the river and on the terraces.

The Maories had registered their right to the greenstone boulder and had not broken it up. Some white men came and broke it up for them, and got Webb, one of the storekeepers (Webb & Murdock), to pack it down the creek. The Maories who had worked the ground about the boulder found out and sued the white men for it. They got possession of the boulder and damages. The case was tried in the Supreme Court at Hokitika.

A deal of permanent work was going on here. Roach and party had a head water race and fluming across the creek and were supplying diggers on the terraces with water for sluicing. The storekeepers on the creek were Webb & Murdock, Taylor, McGlone and Trayner, and R. Gifford. There were no workings very far beyond Maori Point; only two bends beyond it and the ground was very patchy.

1866: We went to the Grey in May and heard that Mr. George Dobson was missing, and at the end of May it was thought he was drowned. But in June, news came down from Nelson about the Maungatapu murders. Burgess, Kelly and Levy had been arrested and Sullivan had confessed to the murder of Dobson, as well as the other murders, and had described the locality where he was buried. Search parties went out under Inspector James and, on the 7$^{th}$ July, Dobson's body was found partly buried in a small creek about 6 miles into the Grey valley track, not far from an iron hut [at a place] now called Dobson Town, about a mile from Brunnerton. A stone monument now (1890) marks the spot close to the railway line. Before the stone was placed there, a tree marked with brass nails marked the spot.

These wretches of murderers were about here for some time. They stopped in a shanty at the Teramakau then went on to the Grey, and no doubt murdered many others as several men were missing at this time and never accounted for.

I saw Burgess and Kelly taken once in Weatherstone's Gully [on the Otago goldfields] in '61 [actually 1862] for some robbery, but Sullivan was the worst of the crowd. I saw him up the Grey, before Dobson's murder, on the track near the iron hut with a double-barrelled gun. I was going up the track with several others and we were all armed. I spoke to him, as he had no swag on, and he said he was shooting pigeons. He said 'Good day' and went off into the bush. I believe if we had been coming down [i.e. carrying gold] he would have stuck us up, as it must have been only a day or two after we saw him that they murdered Dobson out of sheer brutality, as they knew Fox was not like Dobson;[1] no doubt they intended to murder everyone coming down from the diggings.

We were some months about the Greenstone and brought in a head race, and a tail race tunnel to work part of the creek. Dick Sellars was killed by a boulder falling on him from a terrace. We had once a very heavy gale of wind and all the diggers had to leave their huts, on account of trees falling, and take shelter in the stores in the open ground on Maori Point. Sandy Stewart and Yankee Jack's hut was set on fire by a tree crashing through it and a keg of powder went off; they had fortunately just left [the hut] but left their fire burning. This was in the night. Next morning Sandy's gold, which he had in a bag in his trouser's pocket hanging in the hut, was found all melted up and run into the chinks in the floor.

In August '67, Sandy and I went for a month prospecting round Lake Brunner, but got nothing. We had a small terrier who caught us several kiwis which were very good eating. Sandy some years after lost his leg in a claim and his mate was killed. He now lives (1890) near Kumara on Sandy's Terrace, on the road to Hokitika.

---

1 Meaning that Dobson did not look like Fox. The Burgess Gang knew that Fox, a very successful miner and gold buyer, was travelling to Greymouth around that time, carrying a large quantity of gold. Three days later the gang tried unsuccessfully to ambush him again. They went on to murder five travellers on the Maungatapu Track near Nelson, for which three were hanged; the fourth, Sullivan, who had turned informer, had his death sentence commuted to life imprisonment. See Wayne Martin, *Murder on the Maungatapu: A narrative history of the Burgess Gang and their greatest crime*, Canterbury University Press, Christchurch, 2016.

From November '67 to November '68 I was through all the New River[1] district with Welch when a rush took place to the Shed Flat, at the junction of the left and right-hand branches of the Greenstone Creek. A township was started there in a few months and a large police camp on the opposite side of the creek. A great row occurred here between the police and some of the miners. The rioters were fined and imprisoned; one of them, Jack Leathem, got two years for it.

A high terrace on the left-hand side, the Duke of Edinburgh Terrace, was sunk in[2] with fair prospects, and myself and some mates brought in a water race to work the ground and sell water. There were layers of wash in this terrace containing gold for about 200 feet deep, and as more water was got on to it the ground was worked deeper.

A company brought on a larger race from the Big Ohono Creek, some 7 or 8 miles, for sluicing, and the ground is still being worked (1890). The Big Ohono [now renamed the Eastern Hōhono River] falls into Lake Brunner.

In June 1869, Pat May, Morris and myself went over to the Taipo to go to the Seven Mile Creek to prospect for reefs, as it was all reef gold we got there in '64. We went up the Teramakau to the Perounui where we met Charles Russell with a mob of cattle for the Hokitika market. He crossed us over the river with his horses, and we went in to the accommodation house at the Taipo and then up to the Seven Mile, where there were still some men at work.

We stopped some days and went up the creek to get on to the range, but found there was too much snow for us to do anything, so we returned to the accommodation house. We then went up the river Teramakau and crossed into the paddock and got on Howitt's track to the Greenstone. We had a good supply of candles and actually walked all night from the Big Ohono to the Greenstone by candlelight. It was awful work and we were smothered in mud when we got to our hut on the Duke of Edinburgh Terrace.

In August '69, I made another start to find the reefs, with Rowley, Meredith and Wilson. We were still too early, for the snow had not cleared, so we worked part of the Seven Mile. In October we shifted the camp further up the creek; and Roberts and Wallace (Wright's party), who came there for the same purpose, started to cut a track on the right-hand side over the waterfall to get on to the range.

We got on to the reefs in November and all marked out some claims under miner's rights. The same day, the line of reef was pegged out for some

1 The New River/Kaimata, also known then as the Paroa, north of the Taramakau.

2 Prospecting holes were dug into this terrace above Greenstone Creek, opposite the Greenstone township.

distance and several prospecting holes were sunk. Wright's party stopped there for about 12 months and sunk about 60 feet with very good prospects. Men were still working the Seven Mile and have been since '64.

We went to Hokitika in December '69 for about three weeks but there was not the amount of business doing there as previously. The Grey [Greymouth] was getting to be the principal town on the Coast. All the tributaries of the Grey had been and were still turning out large quantities of gold, which clearly proved that the pick and shovel will much upset such decided geological opinions as expressed by Dr. Haast. Returned to Christchurch with Chamberlaine's coach.[1]

On 6th September 1872, I and Day went again over all the West Coast. Left by Cobb & Co. coach. On arrival at Kelly's Creek [near the Seven Mile] we saw the drive [a horizontal tunnel] at work on the eastern face of the Kelly's Ridge reefs, operated by the Pounamu Company. It was quite visible from the creek crossing. We stopped at Alexander's Accommodation House at the junction of the Otira and Teramakau.

Went up the Teramakau to look at a reef, then down the Hokitika road, turning off on a pack road which was made to the Greenstone through the site of the present town of Kumara, which however did not start until '75. Sandy Stewart had a hut on the road, and parties were prospecting about on that side of the Teramakau. Crossed the river in a boat and down the old track to the beach and on to the Grey, then from the Grey to Reefton. The first quartz reefs opened there were in Murray Creek by Kelly, then by Anderson and Shiel on Adam Smith's claim, afterwards called the 'Wealth of Nations'.[2]

1872: In '72 Boatman's Creek[3] reefs were being started, but while we were there in October there was not a great deal doing, as there had not been time to get into proper work in most of the claims.

We found a wonderful difference in having good roads to travel on. We went on to the range at the head of these reefs, and even there, in the slate rocks in the creeks on top of the Victoria Range,[4] we found coarse pieces of alluvial gold, pieces about ½dwt, but not at all payable to work. It merely proves that the main ranges beyond are auriferous.

---

1 At this stage Smart appears to be living in Christchurch again.

2 The 'Wealth of Nations' claim was named after the famous book written by the discoverer's namesake.

3 Boatman's Creek is a tributary of the Inangahua River. It was the site of an earlier, short-lived rush in 1866.

4 The Victoria Range lies between the Maruia and Inangahua rivers.

We got some large kiwis up here with a dog, and some kiwi's eggs. These were very large and one was quite enough for a man's breakfast.

Went some distance along the top of this range towards the Buller; open ground in some places. We then came down to the Inangahua River by following one of the creeks down from the range, and came across some very bad waterfalls which we had to avoid the best way we could. It was bad travelling until we reached the road at the Inangahua River, at the junction of which river and the Buller we found our old friend, Reuben Waite. He had lost all his money and was keeping a small shanty and a ferry boat on the river. We crossed the Inangahua and about two miles further on stopped with Mathew and John Della Vedova, Italians. They were shareholders in the Alpine Quartz Company claim at the Lyell Creek.

Went up to the Lyell on a good road and crossed the Buller in the punt. Found the black sand we started [working] at the point in '63 (all now worked out), a small township, and the old creek and terraces worked about in all directions.[1] Went up to the Alpine Quartz claim on a good pack road. Antonio Zala was manager and he shewed us some splendid specimens of gold quartz from the reef. Went over the workings. This reef is on the range at the head of Irishman's Creek, where such good gold was got in '63. The best patches of gold were got here at the junction of the slate and granite rocks.

From here we went down [the Buller] in one of the horse-boats which now bring up the stores. These boats are towed up by horses. The horses are then put on board the boats and taken back down to Westport. The one we went with had two horses on board and six or eight men pulling [rowing]. We went down in about six hours – different travelling than in our little canoes of '63. The improvements along the riverside by the road and towing tracks shewed that a great deal had been done in eight years.

On arrival at Westport saw John Martin.[2] Went to Waimangaroa and on up Mount Rochfort to have a look at some coal and a small quartz leader which contained very good gold. The coal on Mount Rochfort had not started to be worked. Went over to Addison's Flat and back. A great deal of the shallow ground worked out. Got some very good gold in patches here.

Westport was in a very dejected-looking state as all the easily-got gold was done and very few men were left anywhere. Went back up the Buller road

1 Lyell was a thriving gold-mining town until the 1890s. The town was abandoned around the 1950s; now it has been reclaimed by the bush.

2 John Martin was one of the first storekeepers at the Buller; he arrived in 1861 with Reuben Waite and Isaac Blake.

past the Inangahua; made a small boat, crossed the river and up the Devil's Creek opposite 'Della Vedova's'. Several patches worked out on this creek and terraces. Got gold, very patchy, in coal seam running across a creek. Got 15½oz. Sandstone reefs. Coal full of sulphur; afterwards, on a range on the opposite side of the Buller in a creek, found some good specimens of antimony ore. Another creek at the back of Della Vedova's house. There was a track cut up to this creek and on to the range to take sheep up to a piece of open country. Followed this creek up to its source and got a few colours in several places.

Went to the Grey, Hokitika and back to Christchurch in December 1873.

1890: It is now 26 years since the opening of these Westland Gold Fields, and a very large quantity of gold has been taken from them.[1] All the shallow and easy workings are finished, over a very large area of country. Sluicing claims are now working on most of the old diggings, and in several districts will give steady wages for the next 50 years.

The present race of diggers have lost the enterprise of the old prospectors on the Coast. They are much degenerated and require roads made for them to walk on. They are frightened of being lost in the bush and now just loaf about the townships and rework some of the old ground.

From [the ranges], according to the nature of the wash (granite and slate), all the alluvial gold has come down and been deposited on the marls and sandstones which formed the sea bottom. The belts of metamorphic gold-bearing slates and quartz reefs, which are more or less continuous in and near to this backbone range, are not prospected and some of these reefs appear to dive down to great depths, as outcrops can be seen over 2000 feet below the caps in some of the valleys.

Where the timber has been all cut away a good view can now be obtained of the features of the country, and an idea formed of where the ancient riverbeds and water courses once ran on higher levels.

With this advantage, and the roads, tracks and bridges that are now made, it is surprising that such very little prospecting is yet done to find gold further towards the main backbone ranges of the country.

The End

1 Between 1857 and 1905 the total gold output for Westland and North West Nelson was 6,552,502 ounces (186 tonnes), almost identical to that of the Otago goldfields over the same period. May, *The West Coast Gold Rushes*, p. 519.

***Wattle Bird***

The South Island kōkako, now probably extinct.

## Smart's note on the *Handbook of New Zealand Mines*[1]

Larnach,[2] Minister of Mines, *Handbook of New Zealand Mines*: very incorrect. The information and dates are incorrect. The Little Ohono and Greenstone are one creek; Terapuhi was alive in May[3] and took us across the river to the now called Coal Creek. I and Day and French were the original party who took over provisions [to the Teramakau] in '63 but did not return as stated; and had it not been for us, the gold-field might have not been opened for years, as Mr. Revell was ordered to clear out many times, but my getting Waite to bring down the *Nelson* and Blake the *Mary* started the gold-field straight off.

The greenstone boulder was found as detailed by me and not as stated in that book; nor was it broken up by the Maories. Hunt never saw that part of the Greenstone Creek before he went with the Maories. While we went to the Buller we were not prospecting in the Grey at all. Mr. Revell (see below) had our application for the [gold] reward before we left on the 1st of June 1864, and [in late 1865] I received in Christchurch a letter.[4]

Revell knew that I and French were gone [to the Buller] for a craft for stores, and also that we had left Hunt and two Buller Maories, experienced diggers, to prospect during our absence. We took gold with us to the Buller, and from there sent it to Waite in Nelson with my letter. It was exhibited in Tom Askew's store.[5] Yet this book states that on the 19th of July, Mr. Revell left for Christchurch to report [Hunt's] discovery of gold at the Greenstone simply as if this was the first discovery. On the line of Howitt's track, which goes up the left-hand branch, no gold has been found at all; on the small creeks between the beach and the junction of the Little Ohono and Greenstone Creek a little gold could be got.

This book again states that the Taipo auriferous quartz was first found in 1874, when I now hold the certificate of registration for our claim then dated November 9th 1869. If all information is similar to this in Larnach's book, it is valueless as a true record.

---

1 Transcribed from a note Smart made in the margin on p. 153 of 'The Westland Goldfields'.

2 The author of the *Handbook* was Patrick Galvin. William Larnach as Minister of Mines had commissioned the work and signed the preface.

3 Smart is mistaken here: Tarapuhi died in April 1864.

4 This was the reply he eventually received from the government, rejecting his and French's claim.

5 Tom Askew, a shipping agent and general store owner in Bridge Street, Nelson.

# PART 3

# A smouldering resentment: Smart and the gold reward

*Gold! Gold! Gold! Gold!*
*Bright and yellow, hard and cold,*
*Price of many a crime untold;*
*Gold! Gold! Gold! Gold!*[1]

Smart's purpose in writing 'The Westland Goldfields' was to present his version of events leading to the discovery of a payable goldfield in West Canterbury, and in doing so to justify his claim to be the rightful winner of the government's gold reward. Although this narrative remained unpublished, he had already placed other shorter versions of it before the government and the public, but to no effect. Nevertheless, all these accounts show that he had a case to argue. This chapter draws on the wealth of contemporary resources now available to tell the story of Canterbury's gold reward in more detail, and to fill some of the gaps in Smart's narrative.

As explained in Part 1, the reward was gazetted in September 1861 and advertised regularly until March 1863, when it was withdrawn. Until then, Canterbury prospectors had searched almost exclusively in the east of the province, where there was no gold to speak of, and reward applications were few. But the Dixon/Tainui affair had raised the notion that Canterbury's goldfield would probably be found in the west of the province. It also highlighted problems with the reward's conditions. How was 'the discoverer' to be identified? What was meant by 'reasonably accessible from Christchurch'? West Canterbury could not be further from Christchurch, nor more inaccessible. And how was 'payability' to be proven? The approval of the provincial geologist was required, but even he could offer only an educated guess as to whether a goldfield would pay. At the end of March 1863, the provincial government withdrew the £1000 reward. Instead, it would be 'prepared to treat liberally with any person who has definite information to divulge' – a statement that was even vaguer than that of the reward it replaced.[2] But the change was poorly publicised, and many were under the impression that the £1000 reward still stood.

The key to any such reward was always proof of 'payability'. And the only way to prove a goldfield was payable was to have miners successfully working it, which meant ensuring adequate food supplies. William Smart knew this.

From his arrival at the Hōhonu River in January 1863, his entrepreneurial approach was evident. As he noted in his diary, before leaving for the Buller he sent a letter and gold samples to his Kaiapoi friend Captain Beswick, hoping it would result in a ship being sent around to the Coast with provisions to allow prospectors to return there. Beswick passed on the information in a letter to the provincial secretary,[3] but nothing happened. Smart would later complain that the government's failure to respond delayed West Canterbury's first gold rush by over a year.

Before roads were built, early travellers would trek along the beachline – the easiest and safest route up and down the West Coast. They would carry on their back all the necessaries for the journey rolled into a swag – usually their bedroll. Smart recorded that when they set out to cross to the Coast, each member of his party carried a swag weighing 1 hundredweight (nearly 51 kg).

'McKay and Adamson rounding a bluff below Okarito' [1890s], photographer unknown. Ref: 15272¼. J. Adamson Collection, Alexander Turnbull Library, Wellington, New Zealand.

Smart was away at the remote Lyell diggings for most of 1863, and unaware that in June the provincial government's depot had been established at the mouth of the Grey River near Māwhera pā. The depot's primary purpose was to support the surveyors and track-cutters, but it was also for 'the protection and relief of casual travellers as may be in need of assistance'.[4] Most of these would be prospectors, who were frequently in dire need of provisions. The depot served as general store, post office, meteorological station, vegetable garden and resting place; and as it was near the pā and the port, it soon became the communications and trading hub for that part of the Coast.

But 1863 was a bleak year for the government's West Coast ventures. The pass that Henry Whitcombe 'discovered' proved unsuitable for a road across the Alps; and tragically he drowned while crossing the flooded Taramakau River.[5] Later, depot manager Charles Townsend and two companions (one of whom was the Māori, Solomon Tuapaki), and track-cutter Charlton Howitt, also with two others, all drowned in two separate incidents. Robert Bain and Arthur Dobson, extending Drake's earlier survey, each reported a 'catalogue of disasters'.[6] News from West Canterbury, it seemed, was reduced to reports of deaths, adversity and complaints from disgruntled prospectors.

For most of that year the Hōhonu diggings were almost deserted, except for a few Māori miners and one or two Pākehā such as William Martin, earning a small but steady return. Talk of a gold rush, triggered by the Dixon/Tainui affair, subsided.

These events led to questions being raised in government circles about maintaining its presence in West Canterbury. Thomas Brunner, it seemed, was right: the region was too dangerous and too remote to be of benefit to the province, and there was little point in investing further resources there. But for the time being, the government depot remained open; and in January 1864, William Revell arrived (with his brother John) to take over as manager after Townsend's death in October.

For Smart, the turning point came when he had been at the Lyell Creek diggings on the Buller for nearly a year. As he recorded, in January 1864 Tarapuhi arrived at the Buller River mouth with a party of companions from Māwhera and told Smart that beneath the great greenstone boulder in the riverbed of the upper Hōhonu, Simon Tuakau and his mates had found many small gold nuggets. He also told Smart about the government depot at the Grey, which would supply him with provisions if he returned to prospect at the Hōhonu. Having seen samples of the gold, Smart needed no convincing. His first impressions of the Hōhonu had been confirmed, and now at last there was a secure food supply nearby.

Buller storekeeper Reuben Waite also recorded Tarapuhi's visit: 'a lot of Maories came to the Buller overland from the Grey … They had with them some very nice pieces of gold which they say they have got in the Grey district.'[7] Some said this gold was the first to be found in the area, but of course Ihaia Tainui and others had been digging gold out of the Taramakau and Hōhonu rivers since 1862.

On hearing news of the gold, the enterprising Waite decided to charter a ship to take provisions down to the Grey. Glossing over the fact that the government depot was already operating there, he even claimed the credit for opening the new diggings in the south: 'Two men [Smart and French] started from the Buller to the Grey when they found I was likely to send a vessel down there.'[8] He also wrote to the provincial superintendent, asking what reward was being offered for the discovery of a payable goldfield in West Canterbury.[9] Without waiting for a reply he proceeded with his plans to sail to the Grey, intending to set up his store near the new goldfield, although his departure was delayed for several months.

As he recorded in his diary, in February 1864 Smart left the Buller to travel south with fellow prospector Michael French, Tarapuhi and the others from Māwhera. John Revell noted their arrival at the Grey: 'Tarapuhi says that Mr. Waitt [sic] of the Buller is going to send a vessel round here with stores for a prospecting party under the superintendence of Mr. W. Smart.'[10]

However, when William Revell returned to the depot after a trip to Christchurch at the end of March, he brought the news that the government had now withdrawn the gold bonus. He had been directed to inform all prospectors, and he later declared that he had made the decision known to Smart and French, and also to the newcomer Albert Hunt, who had been a member of the work party completing Howitt's track.[11] The government's change of heart was confirmed in the response Waite received to his earlier letter: 'The Government are not prepared to give any bonus for [the discovery of a goldfield] in that district, as it would be of no benefit to Christchurch!'[12]

It seemed this was the beginning of the end of the government's involvement in West Canterbury. Both of the recent surveys of the West Coast were now completed, and no more were planned. There were even doubts as to whether the Grey depot should be maintained.

As Smart related, the government instructed Revell on several occasions over the next few months to stop selling provisions to the remaining prospectors, other than to supply them with enough to travel north to the Buller or back across the Alps. Fortunately for Smart and French he made an exception for them, as they had helped to move the depot building to a new site out of the reach of floods. Smart saw this concession as significant, for it allowed the pair to carry on prospecting when others had been forced to leave.

Meanwhile, Albert Hunt had visited the Hōhonu diggings and inspected the ground that William Martin's party had been working; and before he and his fellow track-cutters returned to Christchurch to be paid off, he spoke about joining Smart in prospecting up the Greenstone on his return.

In April, Smart and French went out from their camp at the Grey to prospect the neighbouring country, but came back with their Māori mates to their old ground on the Taramakau, at the mouth of the Hōhonu River. There they found Hunt's party had returned from Christchurch and were digging on Smart's former workings, using his old sluice-boxes. When Smart and French returned from another trip up the Taramakau to the Taipo River, they found Hunt working at the Hōhonu alone.

After Hunt's mates left the Coast, he had 'spent nearly all his time at the Maori settlement [Māwhera pā]', according to William Martin. Revell was not supposed to dispense 'continuous charity'; but he could not allow the man to starve, and supplied him with provisions on several occasions. When pressed, Hunt would go out prospecting for a few days, 'then idle [for] several weeks'.[13]

Smart now invited Hunt to join them and prospect higher up the river, but recorded that 'he preferred staying at the Mouth of the Ohono as he was short of flour and there he could buy potatoes from the Maoris'.[14]

In May 1864, after more than two weeks of digging near Maori Point, Smart and French struck pay dirt, having 'followed up the main branch of the Ohono – the Greenstone Creek – where we found payable prospects'. Satisfied with their find, Smart told Hunt that he and French would return to the Buller for supplies, and would be away for about a month.

At the Grey depot, William Revell recorded their comings and goings in his journal:

> May 22. In the evening Messrs Smart and French returned from a prospecting tour up the Teramakau and between the two Hohono rivers [the Greenstone and Little Hōhonu] and report having found gold in payable quantities about the Hohonu Ranges …
> May 23. Supplied Hunt with a fortnight's rations. Was informed by Hunt that his party had made over seven ounces of gold during the eight days they were digging …

Then, on 24 May:

> Received a letter from Smart and French giving a short account accompanied with a rough sketch of the country prospected by them during the last two months [see map on page 92] and of which they speak favourably.[15]

This letter was their formal notification of payable gold in the Greenstone Creek. In it they stated:

> There are many places we prospected up this river that would pay very well for sluicing and in fact we saw a paddock tried of it by 4 men who worked at it easy for eight days, and they got nearly seven ounces of gold, and this was on an indifferent part of the river. About 7 miles up the Ohono … we got prospects of 3–4 grains [of] gold of a shotty description.
>
> Want of provisions compel us to go on to the Buller River to try and get a vessel to come down here at once with stores. We shall return by her and in all probability a great many more will come with us, for, from what we

> have seen, we feel very certain that there is a good payable goldfield in this locality.[16]

Grasping the importance of this news, Revell dashed off a letter to the provincial secretary, forwarding Smart and French's letter and spelling out its implications:

> These gentlemen speak most favorably of the locality prospected as a good payable gold field … as soon as it becomes known that a supply of stores can be obtained here, I am confident that a rush will set in from the Buller … If [so] … I shall take the first opportunity of proceeding to Christchurch with full particulars and await your instructions.[17]

This was the first time a rush to West Canterbury had been formally predicted with such certainty.

On 1 June, when the weather allowed, Smart and French left for the Buller with a large party of Māori. Their purpose was to broadcast the news and make the necessary arrangements for a rush to the Greenstone, namely to round up a large number of miners and charter a ship to transport them and a load of supplies. As Smart recorded in his diary, this would in turn attract more miners, the rush would build, and the goldfield would soon prove payable.

Albert Hunt, having earlier declined Smart's invitation to carry on digging on the Greenstone while he and French were away, now changed his mind. The Māori miners took pity on him, and 'out of good nature' told him where they had found good gold and how to get there. Revell gave him a month's supply of provisions; and Simon and Ihaia Tainui took him in their canoe as far as the Hōhonu and directed him to the diggings at Greenstone Creek, where Smart and French had found their gold.[18]

On 20 June, Hunt returned to the depot and told Revell that in the past two months he had washed 15 ounces of gold, of which he presented just a sample. Hunt applied for a prospecting claim,[19] which Revell accepted and authorised him to peg out the area.[20] Then, having replenished his provisions, Hunt returned to the Greenstone.

Just five days later, Revell received instructions from Christchurch to sell off all his stores and close the depot by August. But having seen samples of Hunt's gold, Revell decided to visit the Greenstone to see the diggings for himself. Writing to the provincial secretary in mid July, he reported that when he arrived he found three parties of Māori and Albert Hunt, all working separate claims, and witnessed Hunt weighing 5 ounces of gold for one day's work. He commented that in contrast to the inefficient methods of the Māori diggers, Hunt had 'worked on the Australian and Otago gold fields and thoroughly understands his business':

> I remained one day on the diggings examining and prospecting about, and from what I have seen and heard I have not the slightest doubt but that the Hohono gold fields, which extend over a large area of country containing

> gold in more or less payable quantities and capable of carrying from 8 to 10,000 people, are now a reality and only await the arrival of some diggers to start them.
>
> Albert Hunt has now in his possession over forty ounces of gold and has made application to me for the prospecting claim, being the first person who has actually obtained the precious metal in a payable quantity.[21]

Revell, who was becoming increasingly aware of the urgency of the situation, decided to go to Christchurch to report to the government in person, and to urge it to reconsider closing the depot. Leaving his brother in charge, he set out, carrying Hunt's letter to the provincial superintendent applying for the gold reward. This letter was printed in the *Lyttelton Times* on 6 August:

> 15 July 1864
> To His Honor, Mr Bealy.
> Sir, I have been on the West Coast since 15 April, prospecting the Hohono, as far as the Perounui, and several gullies running into the Big Hohono river, and found payable gold for several miles on the Hohono. I can get the colour at any place; the prospects are from a half pennyweight to three pennyweights to the dish. I have got about 38 ounces. A party of four might make about £150 per week ... There is room for a large population ... I have been on the Australian and Otago diggings. There are no provisions here; I have been living on 'spuds' most of the time.
> Albert Hunt.

Both Hunt and Revell appear to have forgotten about Smart and French. The pair had been gone for weeks, and who knew when – or if – they'd be back. For Revell it may have been a case of 'out of sight, out of mind'; for Hunt, their absence was to his great advantage.

Meanwhile, having left the Grey at the beginning of June, Smart and French had arrived at the Buller and spoken to storekeeper Isaac Blake. Smart gave him a letter about their discovery, which Blake undertook to deliver to Reuben Waite, then in Nelson. Some weeks later, Waite received Smart's letter and forwarded it to the *Nelson Examiner*, which printed it on 12 July, along with Waite's own embellishments and the editor's comment on his confidence: 'So satisfied is Mr Waite with the above information, that he is now in treaty for chartering the steamer *Nelson* for a trip to the Grey for the purpose of taking with him a large quantity of provisions.'[22]

Waite had no difficulty recruiting miners in Nelson, many of whom were at a loose end. On 18 July the *Nelson* set sail for the Grey with 30 tons of freight, and after calling in at Motueka, Collingwood and the Buller, the number of miners on board had risen to 69. So sure were the steamer's owners that a rush to the Greenstone was imminent that, even before the first voyage departed, a second was planned.[23]

At the Grey, Smart and French's strategy began to take effect. On 21 July, just a week after Hunt had persuaded William Revell he had found payable gold at the Greenstone, Revell reported that the first party of miners recruited by Smart and French had arrived overland from the Buller 'en route to the diggings'; the party of six was 'guided by Mr. French who reports the charter of the schooner *Mary* to bring Mr. Blake's store from the Buller'.[24]

Two days later, John Revell wrote to inform the government that the steamer *Nelson* had arrived, bringing about 70 miners and storekeeper Reuben Waite: 'There is every chance of a heavy rush setting in as there are between 200 and 300 men waiting for the return of the *Nelson* with news of the Hohono [Greenstone] Diggings.'[25]

French and his party soon left the Grey for the Greenstone diggings, with the *Nelson*'s passengers following the next day.[26] The *Mary*, with Smart and 30 passengers on board, having been delayed by the weather, arrived at the same time as the *Nelson* returned on her second voyage.

Waite recalled that most of the local population had gone to the Greenstone, which made the new arrivals all the more anxious to get there. They soon set off. The journey from the Grey to the Greenstone diggings was a 14-kilometre trek south along the beach towards the Taramakau River, then 10 kilometres up the north side of the Taramakau to the mouth of the Hōhonu. From there it was a further 10 kilometres up the riverbed to the diggings.[27] Despite this distance, Greenstone goldfields from this time were frequently referred to as 'the Grey diggings'.

Waite now found himself the only Pākehā left at the Grey among the Māori women.[28] A week later, he recorded, 'some Maoris came down from the diggings and brought with them a sample of about fifty ounces of the finest gold I had ever seen. I was pleased to see it, and purchased it off them.'[29] It seems they had been saving it for his arrival.

They also brought news that events at the mouth of the Hōhonu had taken a violent turn. To protect their ground, some of the established Māori miners tricked the newcomers by blocking the access track to the Greenstone diggings and setting up false diggings down on the Taramakau, where they knew there was little gold. On discovering the deception, the newcomers were furious. They blamed Waite for causing a false rush and stormed back to the Grey and into his store, threatening to hang him and looking to create mayhem.

William Revell reported to the provincial secretary that his brother had been pelted with stones as he crossed the lagoon in his boat. Furthermore, for appearing to have caused a false rush,

> Smart and French have had a narrow escape from being severely handled by a mob of the rowdy diggers who returned to the Grey from the diggings in search of them, but they got away into the bush and concealed themselves for a fortnight, and getting stores from the Depot.[30]

Despite the angry stand-off, a full-scale riot was averted.[31] The blocked track was discovered and opened up, and the Greenstone Creek was rushed. Miners outraged by recent events had intended to leave by the next steamer, but when they saw the quantities of gold being brought down from the diggings, they decided to stay. A month later tensions had eased, the diggers were at work and most were reasonably satisfied with their results.

The diggings expanded at an astonishing rate: the population soon reached nearly 300. A report in the *Lyttelton Times*, dated 2 September, gives a snapshot of the scene during these early weeks of the first West Canterbury gold rush:

> [The men] working on the Greenstone Creek at the present time ... come principally from the Buller and Wakamarina gold fields, induced hither by the glowing accounts forwarded to Nelson and other places by Messrs. Smart and French, who stated in general terms that 'men could make about £20 per week per man'. Hunt, apparently not willing to be outdone, carried his gasconade [boasting] to a higher pitch, and stated that four men could realise £150 per week. Thus the news of the Hohonu has spread in two different directions, and side by side may be seen men who have made all haste to reach this new El Dorado from Christchurch, Dunedin, and Picton.
>
> ...
>
> The natives, however, managed to obtain the best ground that has yet been found. The claim of Simon the Maori has yielded about four pounds weight of gold – some days washing as much as six ounces a day to four men ... Many old miners ... told us that they were making wages or more, signifying in mining parlance about £6 or £7 a week.
>
> There are two stores on the Grey, kept by Messrs. Waite and Blake, from the Buller; a branch store of Mr. Waite's at the mouth of the Hohonu; and a new one on the Greenstone erected by Messrs. Macfarlane.[32]

A month later Revell reported again to the government:

> Numbers of miners still continue to arrive both overland from Christchurch or by steamer from Nelson and there are over 500 miners in the district ... The Greenstone Creek is now fully taken up with claims for a distance of 6 miles up from the junction ... The Maori claim is reported to be the best on the creek and the largest nugget 10oz 16 dwt was taken out of it about a fortnight ago. One or two parties have gone south of the Teramakau prospecting either up the Arahura or the Hokitika but up to date there has been no news from that quarter ... There are about 20 pack horses constantly employed on the beach ... also two bullock drays which make about two trips in the week.[33]

Three men working at an early sluicing operation. Water from a nearby stream was diverted along a flume or water-race onto the face being worked, washing down gravel into sluice-boxes, where the gold was separated and trapped. It was hard and dangerous work: sluicing frequently caused cliffs or terraces to collapse. During the first three years of the gold rushes, 20 percent of sudden deaths on the Coast were due to mining accidents.

'Men sluicing on the West Coast' [1860s], photographer unknown. Ref: 1957.26.57. K. A. Webster Collection, Canterbury Museum.

The gold rush at the Greenstone was proving a huge success, but who could claim the credit for it? This question would be the focus of prolonged and heated debate, as we shall see in the next section.

Meanwhile, Albert Hunt, although apparently doing well on the diggings, had begun to be seen as a troublemaker. As Smart noted in his diary, Hunt had 'persuaded some men to jump some of the Maories' ground', and had even armed himself with a gun to assist proceedings. This clearly angered Smart, as the Māori miners had been kind to Hunt while he and French were away at the Buller.

Revell, too, had changed his mind about the man, having found out that Hunt had been forced off the Greenstone diggings by angry miners, and that in the scuffle he had knocked one of them to the ground with a shovel.[34] In his report to the provincial secretary in late September he wrote:

> A. Hunt the original prospector has been obliged to leave this locality and has returned to Christchurch. He met several of the miners on the road and

> gave them the most flourishing accounts of the diggings, stating that men were taking out 2lbs a day and that there were 2000 ounces in the hands of the police, that he had made his pile of 60 to 100 ounces and was going to Christchurch to claim the reward. Whereas if the truth were known I believe he has never made more than 12 oz since he first began to prospect and that his statements to me from time to time have been nothing more than a tissue of falsehoods and that his statements are in no manner to be relied upon.[35]

Hunt did not return to the Greenstone.

William Smart, on the other hand, remained in the district for several more years, making a fairly decent living from the gold and prospecting further afield. True to his entrepreneurial nature, and drawing on his construction skills, he installed the first of several water-races on the Greenstone diggings to sluice the ground on the Duke of Edinburgh Terrace, and operated it as a business for a short time.

During this period there were some repercussions from his old life back in Kaiapoi.

Just a year after Smart and French first ushered the miners they had recruited onto the Greenstone diggings, more than 500 people were living and working there, including Smart. This early photograph is patchy, but the township can be seen on one side of the creek (foreground), and diggers' huts and tents on the other side, among the mining operations. Conditions were primitive and harsh, but – at first, at least – there was plenty of gold here.

[Greenstone township and diggings, 1860s, photographer unknown], History House, Greymouth.

At the end of 1864, the Major of No. 5 Company of the Canterbury Rifle Volunteers notified the military authorities that their Lieutenant Smart had been absent without leave for more than a year (it was more like two years) and that he would now be stood down and replaced.[36]

More seriously, Smart's former business partner, George Masterton Clark, was called before the Supreme Court in February 1865, a 'prisoner of debt', charged with offences relating to their partnership.[37]

But Smart was far away, and had other matters on his mind.

Finally, the provincial government declared a goldfield in West Canterbury, which was gazetted on 2 March 1865.[38] It was confirmed: Māori title to the region had indeed been extinguished. The Canterbury government had authority to issue miners' licences, collect gold duties, enforce law and order, and sell land. William Revell was appointed goldfields warden and resident magistrate, and a gold escort was organised to carry gold safely across the Alps to Christchurch.

Thousands continued to flood to West Canterbury's goldfields, spreading out from the Hōhonu diggings to the new port town of Hokitika, up the Grey valley and south towards Ross and Okarito. As more goldfields were discovered, still more people arrived. New shanty-towns sprang up almost overnight, populated by miners and the usual hangers-on: storekeepers, tradesmen, hoteliers, grog-shop proprietors, skittle alley and brothel owners, fraudsters and 'loafers'. The gold rushes were proving a rip-roaring success.

But now the government's declaration of a goldfield triggered another rush – one by claimants to its defunct gold reward, hopeful that it would now be resurrected. There were some fresh applicants; others with renewed hope who believed they still had a case; and a few shameless fraudsters.

John Peter Oakes, brother-in-law of Captain Dixon, was first out of the blocks. He wrote to the provincial government on 4 March 1865, claiming that he had commissioned and bankrolled Dixon's expedition to the Taramakau in 1862 when he found gold, and therefore that he (Oakes) was entitled to the reward. The official reply stated curtly that he had no just claim.[39]

Undaunted, Oakes went public. On 21 March, in a letter to the *Lyttelton Times*, he insinuated himself into the centre of the gold reward question as the mastermind behind not only Captain Dixon's discovery but also Albert Hunt's. Again he claimed it was he who 'undertook the task and formed a party and equipped the schooner, *Emerald Isle*, in the month of September 1862, and dispatched the said vessel … in search of gold and sheep country'.[40]

According to William Martin, Captain Dixon had described his 'young gentlemen' passengers as having chartered the vessel to search for grazing land in the west;[41] but Oakes declared their story a pack of lies. Furthermore, he denied that Dixon had entered the Taramakau River mouth in error, as even from the ship's position off the

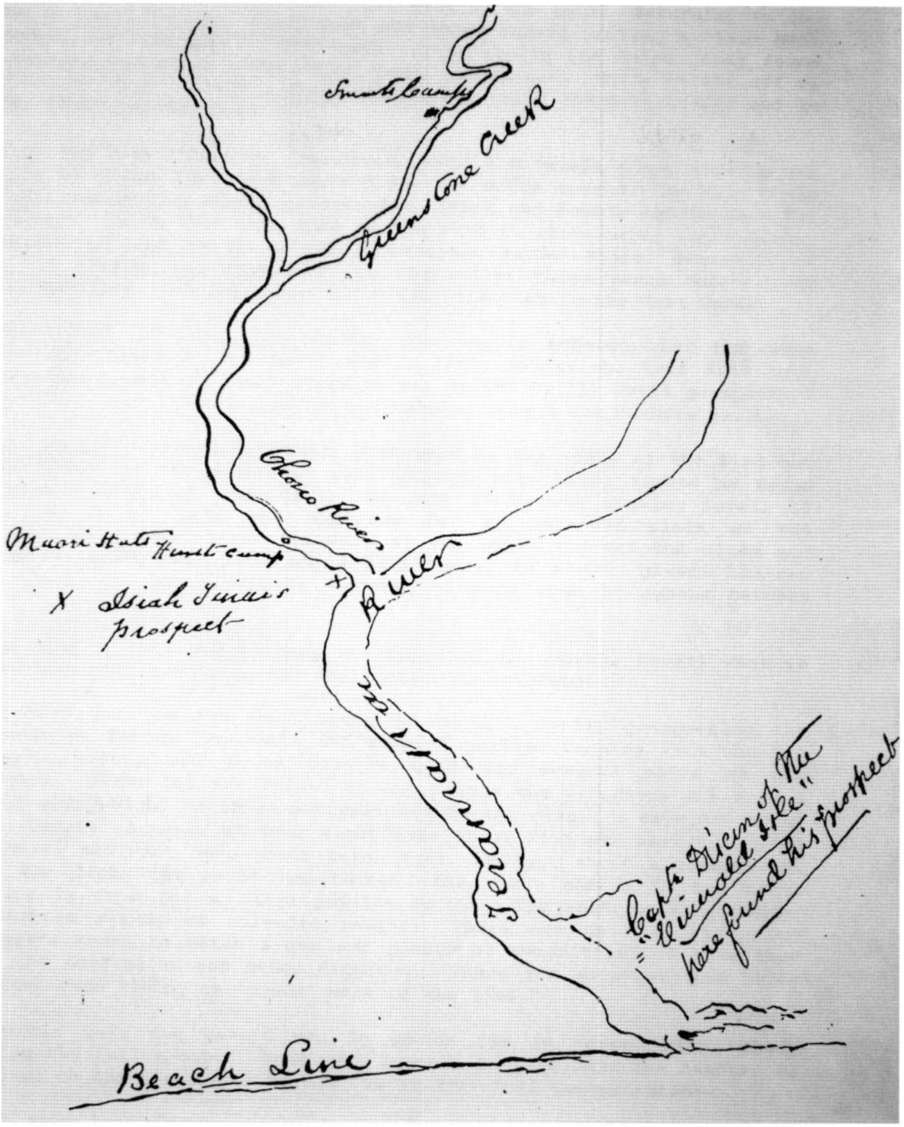


The map shows where four of the five main contenders for the gold reward were positioned on the Taramakau and Hōhonu/Greenstone rivers before the rush began. Near the Taramakau estuary Smart has marked the spot where 'Captain Dixon of the *Emerald Isle* … found his prospect' in 1862. At the mouth of the 'Ohono' [Hōhonu] River are the 'Maori huts', 'Isiah [Ihaia] Tainui's prospect', and 'Hunt's camp'. 'Smart's camp' is located high up the Greenstone Creek, near Māori Point, where he and Michael French first found payable gold. The map does not include the Hokitika River (further south), where Jakob Lauper found gold.

William Smart, ['Map of the Taramakau and Hōhonu/Greenstone gold diggings']. William Smart to the Superintendent of Canterbury, 26 January 1866, in Les Wright, *Greenstone Gold: The first West Coast goldfield 1864–1939*, Department of Conservation, [Hokitika, 198?], p. 17.

coast he had 'judged favourably the auriferous appearance of the country and resolved to enter …'.

Oakes went on to argue that the rush to the Greenstone 18 months later was all his doing: 'It was by means of information laid by me, and claiming the reward [on behalf of Captain Dixon], as well as confirmatory statements through other sources,' that the government sent James Drake to West Canterbury to conduct the first survey expedition in January 1863, and again in December to finish cutting Howitt's track. Oakes claimed that the government had paid Drake 'a ridiculously small sum … and provided no advance supply of provisions,' and that he himself had generously come forward to pay for equipment, provisions and other necessities. Furthermore, 'I also engaged all his men, amongst whom was Mr. Albert Hunt, whom I knew since he was a small boy.'[42]

All his largesse, it seems, was in return for a little gold prospecting, a share of the profits, and ultimately his right to a share of Hunt's gold reward – which was where all this was leading. Oakes failed to present any credible evidence for these scenarios, however, and the government declined to engage with him. This was the first of a number of unsuccessful appeals for the reward by members of the Oakes family over the next 30 years.[43]

Other previous applicants approached the government requesting a reply to their letters and taking the opportunity to restate their claims. Ihaia Tainui wrote in March 1865. In April, Jakob Lauper again put his case to the government, one he had first made in 1863 after finding gold in the Hokitika River during his expedition with Henry Whitcombe; now a major rush had broken out near the site. In May, Albert Hunt and Ihaia Tainui both wrote again to the government requesting a reply to their applications.[44]

William Smart, having received no satisfactory response to his claim in May 1864, also felt compelled to restate his case. In early June he reminded the government that he had applied for the gold reward the previous year, and now submitted his 'Diary of a prospecting tour', which provided more details.[45] The accompanying letter made his point very concisely:

> [It was] through my means the West Coast Goldfields in this Province were first opened. To the above statement Mr. W. H. Revell and Mr. Rochfort, Govt Surveyor, West Coast, can bear testimony. Under these circumstances I beg to put in a claim for the Reward.

French took no further part in the lobbying, but Smart persisted. Two weeks later he wrote again requesting a reply, and took the opportunity to add more information about his prospecting efforts east of the Alps before going to the West Coast.[46]

Finally the provincial government felt obliged to attend to the matter of the reward, hoping to put it to rest. Applicants with the most substantial claims were selected: Ihaia Tainui, Captain Dixon, Jakob Lauper, Albert Hunt, and Smart and French. William Revell, commissioned to assess each application and make a decision, presented his findings in August 1865:

> The claims set up by the Maori, Tainui, and Captain Dixon were made about the same time ... I am informed that these claimants ... never found gold in sufficient quantities to pay themselves or induce others to remain working on the Coast.
>
> These claims were made ... in the year 1863, but since then it does not appear that they took any active steps to prove the capabilities of the West Coast as a payable goldfield ...
>
> The claim of Jacob Louper [sic] comes next and is of a similar nature to the preceding claims. In Jacob Louper's journal and report of that unfortunate expedition with the late Mr. Whitcombe, he states he found a few grains up the Hokitika, but ... he has never made any attempt to return and open up or prove to what extent the gold could be found in that locality ...
>
> The claim set up by Albert Hunt is next to be considered. He first came over to this Coast with [Drake's track-cutting party] at the end of the year 1863 and remained until March 1864, during which time I believe he prospected some of the country along the line of track.
>
> Hunt returned to this Coast again early in April 1864 with three mates and worked on the diggings at the mouth of the Hohono until his mates left as they could not obtain a supply of provisions at the Depot. Hunt remained on the Coast, living with the Maoris and came twice to the store for a small supply of stores. On both occasions he gave me to understand that his mates intended to return with stores; that he was doing very well; had made over forty ounces of gold, and that he could make over two pounds stg (£2) a day prospecting, but only brought one or two ounces each time for sale.
>
> In the early part of July he went to the Greenstone with the Maoris and there got payable gold, about eight ounces which he sold to me at the Depot.
>
> I must here state that no reliance can be placed in his statement, for from circumstances that have come to my knowledge since the opening of the gold field I find that his reports as to prospecting, the rich finds made, together with several other matters connected therewith, have no foundation whatsoever but must have been made for the purpose of claiming the reward if the rush set in or payable gold found.
>
> He was obliged to leave the Coast or else he would have received some severe treatment at the hands of the Otago miners.
>
> I am of the opinion that Messrs French and Smart are the only persons

> who have any claim for the reward for I know of my own knowledge that they prospected both the eastern and western sides of the ranges, as set forth in the Journal kept by Mr. Smart.
>
> In February 1864, they arrived at the Grey from the Buller and remained until the end of May prospecting the country up the Grey, Teramakau and Hohono rivers, and lodged the report and rough sketch of the country bearing the date of 24 May with me. They were then obliged to leave for the Buller to obtain stores as I had already exceeded my instructions relative to supplying rations to prospectors on the Coast.
>
> About the 22nd of July 1864 they returned to the Grey bringing with them over one hundred miners and two storekeepers in the steamer *Nelson* and schooner *Mary*, which were freighted with a large quantity of stores. Since that date, the West Coast gold fields have proved payable and still continue so ...[47]

Although Revell had identified Smart and French as the only worthy contenders, the provincial government 'decided to reject all claimants on the grounds that (a) the reward had been withdrawn before the claims were lodged, and/or (b) the discovery did not warrant the payment of a reward'. And there the matter might have ended.[48]

But the government failed to communicate its decision to the finalists. Soon afterwards Albert Hunt, impatient to gain satisfaction, made a further submission:

> Your petitioner ... believe[s] that to his recherches alone and the publicity which he gave in the matter is mainly due the first exodus of miners to the West Coast ... The Greenstone is a highly profitable and enduring gold field and was discovered under great difficulty and privation.[49]

Others joined in what appears to have been an orchestrated campaign, in which Hunt himself was complicit. The newly launched *West Coast Times* decided to back his cause, clearly unaware of the facts.

> Mr. Albert Hunt, who is well known by most miners as the pioneer of the West Coast, has, as we have been informed, despatched a petition to the Provincial Council, claiming the reward ... It was Mr. Hunt who prospected the Greenstone – a field which has been in high favor by the miners for some considerable time, and was the first miner who made the journey overland from Christchurch to the West Coast. With the exception of Mr. Revell, then Government storekeeper at the Grey, Mr. Hunt was the only European on this coast and for a lengthened period was engaged in prospecting alone and unaided. If enterprise and energy combined with a disregard of hardships and danger entitle Mr. Hunt to the reward, we may say that he has fairly earned it, and trust that he will be successful in his application.[50]

But Revell knew who he was dealing with; only the previous year he had warned the government about Hunt's character.[51]

At this point, a mysterious player entered the scene: apparently John Oakes was not the only one with a fictitious claim to the reward. James Lynch Callahan, 'solicitor', wrote several times to the government on behalf of a 'client', a Mr. John Wait, who claimed to have worked on the diggings with Hunt and was therefore eligible for a portion of his reward. (Hunt had worked on the diggings alone – he had no 'partner'.) His client wished him to proceed immediately to Christchurch in order to bring his case, with 'evidence irrefragable', before the provincial council. The application was rebuffed with the standard response: that the government had abandoned any idea of granting a reward.

Six weeks later, Callahan tried again, repeating his claim. His communication then took an even stranger turn: in the matter of Messrs Hunt and Wait, he 'would take no further proceedings should your Honor secure a Commissionership for me at the West Coast gold field', a position he felt well qualified to perform. Needless to say, his efforts went unrewarded.[52] This looked like the work of a brazen Irish con-man called either Callahan or Lynch (among other aliases), who was operating in New Zealand and Tasmania around this time.

Meanwhile the other hopefuls were still seeking a response from the government. On 29 October, Smart wrote yet again that he had received no reply to his letter, and again restated his case.[53] In November Jakob Lauper also wrote again.

Now Hunt, with backing from 'friends at court', escalated his efforts with a direct approach to the 'Gentlemen of the Provincial Council'. Council member James Hawkes, having received Hunt's petition in November, placed it before the council with his personal endorsement, 'praying that the Council would take the matter into consideration, with a view of giving him the promised reward'. Hawkes argued that, because of information Hunt had sent to the superintendent, people had flocked to West Canterbury; there was 'a general opinion among miners and others that Albert Hunt was entitled to the reward'. However, one independent-minded member observed that Hunt was no more deserving than many others.[54]

At the council meeting on 6 December, Hawkes moved 'that a Select Committee be appointed to inquire into the claim of Albert Hunt'. Committee members included Hawkes, and the two new council members representing West Canterbury, Evan Prosser and Edmund Barff, who were also supporters of Hunt. However, in a bid for fairness, the council voted that the committee look at all five main applications.[55] Hawkes, knowing that Hunt was in Christchurch, successfully added an amendment giving the committee the power to call 'persons and papers'.

The select committee's report (now lost) was presented to the council on 21 December and discussed in early January 1866. Mr Barff declared that Hunt deserved some consideration for his discovery of the goldfield; he had been induced to prospect in West Canterbury by the government's advertisements of the gold reward. Hunt's supporters argued that if hard work, lack of clothing, food and comforts, living in a bush shack and

## The Māori gold miners

In spite of all the drama surrounding Pākehā claims and counterclaims, it is obvious that Māori were first to find gold on the West Coast – initially at the Buller sites in Nelson province, and then in West Canterbury. Ihaia Tainui's story shows just how far Māori were ahead of the game, bringing the knowledge of gold digging from the north, and beginning to work the ground at the Hōhonu well before any Pākehā arrived.

Furthermore, they played a crucial role in opening up the goldfields by showing early Pākehā prospectors on to the gold. It was Tainui who, on his return from Christchurch after presenting his gold reward claim, met Smart's party on the Taramakau and took them to where he and his mates had found gold on the Hōhonu. It is clear, too, that the Māori diggers directed Martin, and later Hunt, to where they were getting gold in the same area.

So why did the Māori miners – with the exception of Tainui – make no serious attempts to claim the gold reward? In spite of being the real gold-mining pioneers in West Canterbury, they may have considered themselves ineligible, or felt intimidated by Pākehā officialdom. Tainui, on the other hand, was literate and had the confidence and mana to approach the government, make his claim and persist with it. In December 1863, for example, obviously doing well on the diggings, he wrote to the provincial secretary in Christchurch seeking clarification about the amount of gold that must be discovered to entitle a person to claim the reward.[56] However, the government's rejection of his applications may have deterred others; only one other Māori claim for the reward was reported, but this disappeared without trace.[57]

Nor did any of the Māori prospectors appear interested in becoming entrepreneurs in a full-scale rush. There were probably several reasons for this. Tainui had expressed his concerns about an earlier rush to the first Buller diggings, fearing the evils that an influx of Pākehā miners could bring. He also believed Pākehā were unsuited to the harsh and dangerous conditions on the Coast: 'This place is not fit for Europeans, there's too much water.'[58] This may have been motivated by his desire to keep out the competition, but there was ample evidence to back his concerns.

It appears that most Māori miners were content to make a steady living from the gold. When Charlton Howitt looked around the Hōhonu diggings in late 1862, he saw gold 'everywhere [but] scarcely enough to satisfy white men, though the Maoris there (about 18) seem satisfied'.[59] The Māori diggers were hardy and well adapted to the conditions, and at the Hōhonu they not only had the best ground and the skills to work it, but were canny about getting the best price for their gold.

Although their success was often downplayed by the press, and some of their methods may have puzzled European observers, in those early days Māori miners engaged in the gold-digging business on their own terms.

enduring the perpetual rains of the West Coast were the prerequisite to the reward, then Hunt had surely qualified. Not only this, he had proved in 1864 the presence of gold in payable quantities by his own winnings. It was through his efforts that population had flocked to the West Coast; and had not the amount of gold extracted in West Canterbury leapt from 1463 ounces to 289,897 ounces between 1864 and 1865?[60] Furthermore, they claimed, opinion among the mining population was that Hunt was the only man entitled to any reward for the discovery of the West Coast goldfield.[61]

Then the bargaining began. Barff moved that the superintendent place upon the estimates £1000, the original amount of the reward, to be given to Albert Hunt as a bonus. But the provincial secretary reminded the meeting that the original £1000 gold reward had been withdrawn in March 1863, and Hunt did not find gold until more than a year later.

Hawkes agreed. While Hunt had no legal claim to the reward, 'a sum might be given, however, to him in consideration of the trouble he had taken'. Other members warned against the precedent this would set; another said the money should be given not as a 'gold reward' but as a gratuity for services rendered.

Barff withdrew the motion. The provincial secretary recommended £200 as ample recognition of Hunt's services. His supporters argued that this would hardly cover his expenses. Barff made a counter offer of £500. Hawkes responded with an offer of £250. The vote was lost 19 to 4. The campaign appeared to have foundered.

Two weeks later Hawkes tried again, moving that £200 be paid to Hunt as an acknowledgement of his 'services to the province in exploring and developing the resources of its western portion'. Perhaps wearied by the pleadings, the council passed the motion. Barff and his colleagues were well pleased – their persistence had paid off handsomely.[62]

Back on the Coast, Barff boasted about the outcome. 'Hunt,' he crowed, 'had on all sides the credit of having not only discovered, but developed the Goldfield.'[63] William Revell would have shaken his head in disbelief, having rejected Hunt's case the previous year in his official review of claims; but his report appeared to have played no part in the recent debate.

Needless to say, there was strong reaction from a number of quarters; and none stronger than William Smart's. He was heartily annoyed, and immediately wrote to the government, repeating his claim and enclosing another sketch map of the diggings (see page 123). Hunt was not entitled to the reward, he argued, 'as he did not find [the payable goldfield], for I had told him and the Maoris that there was payable gold in the Greenstone Creek'.[64] The provincial secretary replied that Smart's previous correspondence had been 'laid before' a select committee, which decided that he had 'failed to substantiate' his claim to any reward.[65]

Others, though, agreed with Smart. Historian W. A. Taylor, writing in 1952, didn't mince his words:

> The notorious Albert Hunt, in 1864, dumped himself alongside the claim

> of the Maoris at the Hohonu and claimed the award. Mr. W. H. Revell, the Gold Fields Warden, at first backed up Albert Hunt's claim, but later he found out the truth and his written reports ... are decidedly hostile.[66]

William Martin was also clear on the subject. Hunt didn't deserve the credit or the reward for finding a payable goldfield: he had been shown onto the Greenstone diggings by the Māori miners. He added: 'before the rush set in no less than three parties had found gold [there], our party was the fourth'.[67] But he also stated that, having gone 'as directed' to the diggings, Hunt 'returned to Mr. Revell [at the depot] with an extraordinary fine sample of gold which led to the rush'.[68]

This last statement is of course wrong; Hunt's find did *not* lead to the rush. But it goes to the heart of the matter, which was all about timing, and the false attribution of cause and effect.

Once Smart and French had identified a payable goldfield on the Greenstone and gone north to make arrangements for a rush, Hunt made good use of their absence. When Revell visited the Greenstone in July he thought it showed every sign of being a rich goldfield, and Hunt was able to sway him into believing, for a time, that *he* was the one who first found payable gold there. At virtually the same time, the miners and provisions organised by Smart and French arrived and the rush began. For Hunt it was a perfect coincidence: for it was he who was seen to have triggered the rush. Choosing to remain silent on the role that Smart and French had played, he took the credit until his story became the accepted truth, written into the history books.

As his narrative suggests, Smart never forgave the government; nor did he forget. But he had to accept the injustice and get on with his life – and that, at least for the next 20 years, is what he did.

After the official declaration of the West Canterbury goldfield in 1865, Smart remained on the Coast and continued to make his living from digging gold. He prospected at various other sites, but the Greenstone was his main area of interest. French stayed too, and opened a store there.[69]

In 1869 Smart sold his water-race business at Greenstone Creek for £150. It appears that around this time he gave up full-time prospecting and his life on the Coast and returned to Christchurch. And he married. It seems strange to think of William Smart as having a settled domestic life, but in 1874, at the age of 44, he married Ellen Deighton, a widow of Kaiapoi, and the couple made their home at Grantchester Cottage on Ferry Road. There were no children from this marriage, but some of Ellen's children by her first husband were living nearby, along with their own families.[70]

Now back in town, Smart had to make a living. He applied for jobs as clerk of works for the Lyttelton Harbour Board and in the government's Public Works Department (whether he was successful is not known); he built stone culverts and retaining walls for the new Mainland Railway; and for a time he went back into business as an architect

In 1866, just three years after Smart sketched the same view (see pages 70–71), the port town of Greymouth had become the commercial hub of numerous outlying goldfields. Wereta Tainui had succeeded his brother Tarapuhi as Poutini Ngāi Tahu chief. Māwhera pā still existed, but much of its native reserve land was gradually taken out of the hapū's control, surveyed for town sections and built on by settlers. Finally, as the Pākehā community became more and more dominant, the Poutini Ngāi Tahu hapū moved its headquarters south to the pā at Arahura.

James F. Barke, 'Grey River and Gorge, and east part of the town of Greymouth, Westland, NZ'. De Gruchy & Leigh, Lithographers, Melbourne, 1866. Ref: B-080-017. Alexander Turnbull Library, Wellington, New Zealand, http://natlib.govt.nz/records/22509969.

and contractor, building and renovating houses.[71] His wife opened a day school for young ladies at their home in Ferry Road.[72]

Smart's interest in prospecting and mining persisted, and on several occasions he went back to the West Coast to visit his old haunts, see old friends and do a bit of gold digging. From Christchurch, he began developing these interests more formally by involving himself in various mining enterprises as a mine owner and manager. In 1870 he started the Alpine Quartz Mining Company, one of the companies formed to work the ground on the Taipo Reef at Kelly's Creek, which he had discovered in 1864.[73]

By 1872, however, he had turned his attention to other minerals that were related to his work in the construction industry. In 1872 he opened the White Rock Quarries near Ashley Gorge, west of Rangiora, which supplied 'splendid building stone, now at greatly reduced rates to the people of Christchurch'.[74] Five years later, he opened a lime kiln in the same district, teaming up again with George Day, the brother of his prospecting friend, Robert Day. He developed this business into the Whitecliffs Marble, Lime Quarry and Kiln in partnership with Leonard Stace. However, it appears to have been a relatively short-lived venture: in 1880 he and Stace were before the Supreme Court for debt, and the business was sold.[75]

Smart's main focus, however, was on coal mining in the Malvern Hills.[76] He set up, owned and/or managed several coal mines there, including the Whitecliff, St Helen's, Rockwood and Hartley mines – with mixed success. He maintained his involvement in coal mines almost until the day he died. But prospecting was still his passion, and in 1883 he became a foundation shareholder and director (and later managing director) of the Premier Prospecting Company. The company was granted several prospecting licences for copper and other minerals in the Malvern Hills, and prospected for gold in the hills of the Rakaia Valley.[77] By 1884, however, the company had sunk without trace.

In addition to these new business ventures, Smart presented himself as a consultant – 'a certificated mining engineer'. He certainly had knowledge of geology, a good eye for gold country, and experience in alluvial mining techniques; but when and where he would have 'qualified' is not known. He gained quite a reputation, and his name was used to lend credibility to company share floats and prospectuses; for example, shares in land for sale to investors in the Malvern coal-mining district could be 'purchased on the recommendation of Mr. Smart the well-known coal and mineral expert'.[78]

In business, then as now, it was a common ploy to pander to the ignorance, optimism and greed of potential investors. So-called 'experts' could produce the 'evidence' that entrepreneurs wanted to hear, and had paid for. On more than one occasion, when Smart advised in areas where he had little experience, his credentials were called into question. When acting as manager and mining consultant for the Glenroy Coal Company, he reported to a meeting of the directors that their coal mine in the Malvern Hills appeared, in his expert opinion, to be 'all that could be desired'. This contradicted the opinion of the former chairman and some of the directors, who had declared the coal field to be 'bogus'. His critics were justified: 18 months later the liquidators were called in.[79]

In 1890, when a tin-mining company hired Smart to assess prospects on Stewart Island, he declared there was not only tin but gold in the company's claim. 'Mr. Smart, a mining engineer from Christchurch who is acquainted with the tin mines of Cornwall ... has no doubts that these loads are highly payable ... He is sanguine that his report will result in the matter being taken up by the capitalists in London.'[80] However, although small amounts of tin were found over a wide area, the main tin rush that had begun two years earlier was already over and the London capitalists stayed away.

The most direct challenge came when Smart endorsed the 'excellent prospects' for the Neptune Gold Mining and Sluicing Company, which planned to establish a beach-mining operation just north of Greymouth. Its grand prospectus, printed in newspapers around New Zealand, included a technical report by William Smart, 'Certificated Mining Engineer'. One highly sceptical reader of the *Grey River Argus* dared to question the credentials of 'this well-known mining expert':

> Mr. Smart, I take it that as we are writing about goldmining, that you mean that you are a certificated *gold* mining engineer. If so, tell us from whom,

> how, where, when, and for what you got that certificate. Who dubbed you 'Mining Expert'?[81]

The writer declared that his mission was to expose both 'this bogus company', which had failed to even secure a mining lease for its beach operation, and 'the celebrated Mr. Smart, who tried to float that remarkable claim, better known as Neptune No. 2 and which was as mythical as the man in the moon'.[82]

In 1889 Smart did, in fact, become certificated – not as a mining engineer but as a mine manager. The government certificate was available to anyone of good repute who had at least one year's work experience as an underground mine manager – and 10 shillings.[83] Not that it made any difference to his practice. A government mines inspector visited one of Smart's mines in the Malvern Hills in 1895. Smart had reported that it contained a coal seam 6½ feet (2 metres) thick, but on investigation, the inspector noted that Smart 'failed to show me any coal seam of greater thickness than a paltry few inches ... therefore good money is being foolishly spent in drilling one tunnel after another'.[84]

To the last, Smart was embellishing his credentials. In 1899, just a year before he died, a coal mine in the Malvern Hills engaged him to direct prospecting work: 'Mr. W. Smart, mining engineer, who has considerable experience in Great Britain ...'[85] Smart's determination to succeed in the mining industry never flagged. His obituary described him as 'a man who had a great deal of faith in the mineral resources of Canterbury, and never ceased to believe that at some time very large deposits of coal and other minerals would be found in the Malvern Hills'.[86]

Overconfidence and misrepresentation seem to have been a pattern in Smart's business life. His credentials were never officially challenged; however, no evidence has been found of his claimed experience in the Cornish tin mines, for example, or of where and when he qualified as a 'mining engineer'.

There were of course consequences to his behaviour. His first visit to the bankruptcy court, after the failure of one of his coal-mining companies, was in 1885. He was discharged, but was back again two years later. He related his tale of woe to the judge: his financial difficulties were due to the Canterbury government contracting his company for only a fraction of the coal he had initially anticipated, and then there was the bad weather, and the expenses of opening up the mine. Furthermore his partner, Mr Gundrey, had quit the business; Smart paid him £1 for his share of the assets, and took on the burden of the liabilities. Given his earlier partnership with George Clark, it seemed the tables were now being turned on Smart.[87]

In the midst of Smart's troubles, the distant past came back to haunt him. More than 20 years after Albert Hunt extracted the 'gold bonus' from the provincial government, Smart's old resentments were dredged up once again. The provocation was a series of articles in the *Weekly Press* by a well-known Canterbury journalist, Frederick Rollitt, a regular columnist who wrote under the nom de plume 'the Warrigal'.[88]

In 1887 Rollitt began a travelogue about his visit to the West Coast goldfields, confessing to 'a weakness for rambling about these old digging-places'. In his first article, 'Kumara to the Greenstone', printed at the end of September, he mused about the adventures and ordeals of the West Coast's early explorers, Brunner and Heaphy, Whitcombe and Lauper, and the first prospectors to find gold in West Canterbury. An old miner had guided Rollitt to the diggings on Greenstone Creek. Looking up to the peaks beyond the river he observed:

> There's Mt. Smart and Mt. French, they are named after two of the first prospectors: Smart and French claimed the bonus for the first gold, but Hunt got it. Hunt the prospector, whom every miner knew in those days, he was a grand bushman, and nothing could stop him, he knew the country better than the Maoris did.[89]

Smart immediately shot off a letter to the editor of the *Press.* Under the headline 'First discoverer of gold in Westland', he restated publicly that it was he and French who engineered the rush of 1864, thereby proving that the Greenstone was the first payable goldfield in West Canterbury. They had, on at least two occasions, provided evidence of this to the provincial government, and each time been turned down. Smart ended with an offer to provide, yet again, evidence for his rightful claim to the gold reward: 'I have my original diary, and shall be willing to give you a copy of it if you think it worth publishing.'[90]

Less than two weeks later, Rollitt continued his travelogue with 'Memories of the Grey Valley', in which he described the earliest beginnings of the Canterbury gold rushes. Whether or not he had read Smart's recent letter to the editor is hard to know. He did, however, praise Smart and French, 'those daring adventurers', and acknowledged their place among

> the first of the gallant prospectors; they were successful in their object for I find that in the year 1864 they wrote to Mr. Waite, of Nelson, saying that everywhere they dug they could find the colour of gold. The news this letter contained spread abroad and tempted seventy adventurous miners to leave civilization [to try the new goldfield].[91]

Rollitt took the story no further, but he went on to complain that there was still no 'consecutive history of the early gold discoveries on the West Coast'. These days, he said, the memories of old miners were hazy, and even the official *Handbook of New Zealand Mines,* published that year, fell well short.

On this point, at least, Smart heartily agreed with him. In his eyes the *Handbook* served only to perpetuate the injustice that he and French had suffered with regard to

Hunt's fraud. Smart determined to set the record straight in the pages of the *Press*, this time with a longer letter to the editor. In this disjointed and somewhat inconclusive summary of events, extracted from his diary, he once again attempted to explain the strategy he and French had used to trigger the rush at Greenstone Creek, and to argue that it was they, not Hunt, who deserved official recognition for it, and preferably some cash as well.[92]

Perhaps the editor of the *Press* took up Smart's earlier offer to make his diary available for publication, for soon afterwards Smart began work on 'The Westland Goldfields', his last and most comprehensive version of the story. He used what he had to hand – namely a ledger book – and cut sketches from his diary and pasted them into his handwritten text.

As noted, the manuscript has the appearance of being intended for publication, with its formal title, introduction and (possibly incomplete) conclusion. He worked on it until 1890, making annotations, updating and expanding the text, and venting even more spleen on Albert Hunt and the Canterbury government in the process. The act of writing may have opened up old wounds, but perhaps it was a welcome diversion, allowing him to relive his heyday on the goldfields during those eventful times.

Smart's writing reveals many of his admirable qualities, including that he was energetic, hard-working and versatile. This small, remote colony was a place where a restless young man could reinvent himself and forge a number of careers out of very little. Smart certainly tried many different roles: builder, architect, sheep farmer, soldier, racehorse owner and breeder, prospector and miner, explorer, storekeeper, amateur medic, gold buyer, entrepreneur, company director, geologist, mine manager, writer and artist.

But as we have seen, he had his frailties. After the freedom and success he had enjoyed on the West Coast, the more formal, regulated society of Victorian Christchurch seemed not to suit him. Many of his business ventures ended in failure. And although he had suffered at the hands of a fraudster, some of his own dealings came very close to being fraudulent.

In 1895, Smart made his final bid for recognition in letters to the Minister of Mines and to Premier Richard Seddon. After re-telling his story he made a further case for compensation, this time not for the £1000, but a free pass on the railways. But still he failed to make any impression.[93]

Smart died in 1900 – no doubt a disappointed man. He had not touched 'The Westland Goldfields' in 10 years, and while he was alive it never saw the light of day.

Smart believed that 'The Westland Goldfields' was a factual historical record: he had been there, after all, and his account must therefore be true and authoritative. In reality, of course, it presents just a single viewpoint, and one that is not always accurate or free of bias; and given the time, the place and the conditions it was written under, it could never hope to be comprehensive. Nevertheless, Smart invested considerable time, skill and effort in compiling his illustrated text.

Few other gold miners of the period are known to have kept a diary, and even fewer worked theirs into a larger narrative. This makes Smart's account unique. Unlike other contemporary narratives of the early goldfields, it is personal and immediate, closely concerned with day-to-day events and observations.

Smart may have been motivated by grievance and a desire to set the record straight, but incidentally he has given us a lively social history of the pre-gold rush West Coast, with an insider's view of life on the diggings. Furthermore, the latter part of his narrative describes the rapid and dramatic changes on the Coast as a result of the rushes and their gradual decline.

Smart also gives us a unique insight into his own transformation from a young urban Englishman into a New Zealander completely at home in the bush, living rough and working comfortably alongside Māori. It was more than the lure of gold and the promise of wealth that drove him. On the West Coast, far from civilisation, it was the life itself that he loved: the freedom and adventure, the challenge and satisfaction. Seen in this light, his narrative gives clues to an emergent Pākehā male identity – the 'man alone' and 'good keen man' myths now embedded in our national (and especially West Coast) character.

In the end, Smart's campaign for the gold reward was unsuccessful. But 'The Westland Goldfields' delivers ample evidence to back his claim; and that evidence has been corroborated through the close examination of other accounts and relevant documents. Even the much maligned *Handbook of New Zealand Mines* acknowledged that Smart and French had been the catalyst for the Greenstone rush: 'It was probably in consequence of [their initiatives] that the first large party of diggers proceeded to the Grey.'[94]

But once Albert Hunt had gained his prize, people stopped listening to William Smart. Hunt had stolen his place in our history as the 'heroic' prospector responsible for triggering the West Canterbury gold rushes.

## Epilogue

In the 1960s the New Zealand Historic Places Trust placed a commemorative plaque beside Greenstone Creek.

This tribute to Albert Hunt in bronze and granite shows how durable his fraud has been.

Photograph: Hilary Low.

William Smart would not have been amused. He may have taken a little comfort from the naming of Mt Smart and Mt French – acknowledgement that the pair deserved some recognition. And some might say that these mountains are a grander and more enduring memorial than any plaque on a lonely riverbed. But perhaps, as a consequence of this investigation into Smart's grievances, those responsible for the plaque on the banks of the Greenstone River/Hokonui may consider changing its wording.

## Notes

### Abbreviations

ANZ Archives New Zealand
ATL Alexander Turnbull Library, National Library of New Zealand, Wellington
ICPS Inward Correspondence of the Provincial Secretary (Canterbury)
OLPS Outward Letterbook of the Provincial Secretary (Canterbury)

1 T. Hood, quoted in Alfred Saunders, *Tales of a Pioneer: Episodes in the life of Alfred Saunders, selected and arranged by his two youngest daughters*, Isitt-Wildey, Christchurch, 1927, p. 83.
2 Minutes of the Executive Council of the Province of Canterbury, 31 March 1863, p. 75, CAAR 134060 CH287 CP372 R22397709, ANZ, Christchurch.
3 William Smart, 'First discoverer of gold in Westland', *Press*, 4 October 1887, 5.
4 Journal of Proceedings of the Canterbury Provincial Council, Session XX, 1863, p. 185.
5 Ironically, after his companion, Jakob Lauper, found gold in the Hokitika River, the £1000 gold reward that Whitcombe had imagined himself claiming was paid instead to his widow and six children as compensation for his death. Hilary Low, *Pushing His Luck: Report of the expedition and death of Henry Whitcombe, by Jakob Lauper*, Canterbury University Press, Christchurch, 2010, p. 81.
6 Philip Ross May, *The West Coast Gold Rushes*, Pegasus, Christchurch, 1962, p. 95.
7 Reuben Waite, *Narrative of the Discovery of the West Coast Gold-Fields*, W. H. L. Leech (ed.), J. Hounsell, Nelson, 1869, p. 10.
8 Waite, *Narrative of the Discovery*, p. 11.
9 Reuben Waite to Provincial Secretary, 4 February 1864, ICPS 466/1864, ANZ, Christchurch.
10 J. C. Revell, 'Copy of journal, Prov. Govt. depot', 16 February 1864, ICPS 717/1864; ICPS 718/1864, ANZ, Christchurch.
11 W. H. Revell, 'Correspondence relative to claims made by Albert Hunt and others for the reward offered for the discovery of a payable gold field in the province of Canterbury', 11 August 1865, Canterbury Provincial Council, Papers and Returns, CAAR CH287 20410 Box CP 618 PPC 9, ANZ, Christchurch.
12 Waite, *Narrative of the Discovery*, p. 10.
13 William Martin, in Julia Bradshaw (ed.), *The Diggers' Story: Accounts of the West Coast gold rushes* (An annotated and illustrated new edition of the classic publication, compiled by Carl Pfaff and published by the West Coasters' Association in 1914), Canterbury University Press, Christchurch, 2014, p. 40.
14 William Smart, 'Diary of a prospecting tour to the West Coast, from 1862 to 1865', 1865, CAAR CH287 19936 ICPS 790(1)/1865, ANZ, Christchurch.
15 W. H. Revell, 'Grey depot journal', May 1864, ICPS 1894 (1)/1864, ANZ, Christchurch.
16 William Smart and Michael French to William Revell, 24 May 1864, ICPS 1895(1)/1864, ANZ, Christchurch.
17 W. H. Revell to Provincial Secretary, 25 May 1864, ICPS 1895(2)/1864, ANZ, Christchurch.
18 Martin, in Bradshaw (ed.), *The Diggers' Story*, pp. 39–40.
19 A prospecting or prospector's claim was the first claim in a new locality. The finder was allowed to peg out a larger area than normal, usually double, as a bonus.
20 J. C. Revell, 'Journal of the Grey Store, 30 May to 30 June 1864', ICPS 1998/1864, ANZ, Christchurch.
21 W. H. Revell to Provincial Secretary, ICPS 1999 (1)/1864; 1999(2)/1864, ANZ, Christchurch.
22 *Nelson Examiner & New Zealand Chronicle*, 12 July 1864, 5.
23 'Gold diggings at the Grey', *Nelson Examiner & New Zealand Chronicle*, 12 July 1864, 5; 'Shipping intelligence', 19 July 1864, 2; 'Shipping intelligence', *Nelson Examiner & New Zealand Chronicle*, 16 July 1864, 2. The steamer was owned by the Nelson & Marlborough Coast Steam Navigation Company Ltd.
24 J. C. Revell, 'Journal of labor, events and occurrences at the Prov. Gov. depot, West Coast, 1 July to 31 July 1864', ICPS 2141 (1)/1864, ANZ, Christchurch.
25 J. C. Revell to Provincial Secretary, 23 July 1864, ICPS 2141(2)/1864, ANZ, Christchurch.
26 J. C. Revell, 'Journal of labor, events and occurrences', ICPS 2141(1)/1864, ANZ, Christchurch.
27 W. H. Revell to Provincial Secretary, 15 July 1864, ICPS 1999 (1)/1864, ANZ, Christchurch. At first, supplies were brought to the diggings on foot; later, once the rush was in full swing, they were carried

by packhorses along the beach, then taken up the river in canoes.

28 Waite, *Narrative of the Discovery*, p. 13.

29 Ibid.

30 W. H. Revell to Provincial Secretary, 24 August 1864, ICPS 2238/1864, ANZ, Christchurch.

31 Waite, *Narrative of the Discovery*, pp. 13–15.

32 'The Hohonu diggings', *Lyttelton Times*, 17 September 1864, 5.

33 W. H. Revell to Provincial Secretary, 28 September 1864, ICPS 2040/1864, ANZ, Christchurch.

34 Patrick Galvin, *Handbook of New Zealand Mines*, New Zealand Mines Department; Government Printer, Wellington, 1887, p. 119.

35 W. H. Revell, Report to Provincial Secretary, 28 September 1862, ICPS 2040/1864, ANZ, Christchurch.

36 Thomas Woollaston White, Major No. 5 Company, Canterbury Rifle Volunteers, to Deputy Adjutant General, Auckland, 'Election of officers', 13 December 1864, DAG 1864/3200, ANZ, Wellington.

37 'In the Supreme Court of New Zealand', *Lyttelton Times*, 14 February 1865, 7.

38 *Lyttelton Times*, 4 March 1865, 5.

39 Canterbury Provincial Secretary to John Oakes, OLPS 171/1865, ANZ, Christchurch.

40 John P. Oakes, 'Discovery of gold on the West Coast: To the editor of the *Lyttelton Times*', *Lyttelton Times*, 21 March 1865, 5.

41 William Martin, 'A pioneer's reminiscences', [c. 1913], MS206, Hocken Library, University of Otago, pp. 1–2.

42 John P. Oakes, 'Discovery of gold on the West Coast: To the editor of the *Lyttelton Times*', *Lyttelton Times*, 21 March 1865, 5. Both Sherrin and Drake had formal government contracts to finish Howitt's track, each overseen by a government engineer. Drake's party was supplied from the Grey depot standard provisions for a normal government survey expedition. Thornton to Provincial Secretary, 'Continuing work on the West Coast road via the Taramakau Saddle', 11 January 1864, ICPS 96/1864, ANZ, Christchurch; W. H. Revell, 'Journal of the Grey River store', 23 February 1864, ICPS 590/1864, ANZ, Christchurch.

43 Late claimants for the reward continued to be presented; all were rejected. The most persistent were members of the Oakes family. John Oakes petitioned the government again in 1874, this time with a new and unsubstantiated story that the *Emerald Isle* visited the West Coast in 1857 and found gold. In the 1890s his brother petitioned the House of Representatives for recognition of the 1857 voyage and the Oakes brothers' discovery of both gold and coal. None of the family's claims has been proven to be anything other than fabrications. See May, *The West Coast Gold Rushes*, pp. 520–27.

44 Ihaia Tainui to the Provincial Secretary, 24 March 1865, OLPS 272/1865, ANZ, Christchurch; Revell, 'Correspondence relative to claims'.

45 Smart, 'Diary of a prospecting tour'; Revell, 'Correspondence relative to claims'.

46 William Smart to Provincial Secretary, 5 June 1865, in Revell, 'Correspondence relative to claims'.

47 Revell, 'Correspondence relative to claims'.

48 May, *The West Coast Gold Rushes*, p. 523.

49 Albert Hunt, 'Petition relating to the reward offered for gold discovery', Session 24, 23 November 1865, CAAR 20410, CH287 CP616/d PPC2, ANZ, Christchurch.

50 *West Coast Times*, 21 September 1865, 2.

51 W. H. Revell to Provincial Secretary, 28 September 1864, ICPS 2040/1864, ANZ, Christchurch.

52 James Callahan, Letters to Superintendent, ICPS 1668/1866, ANZ, Christchurch.

53 Provincial Secretary to William Smart, OLPS 804/1865, ANZ, Christchurch; Revell, 'Correspondence relative to claims'.

54 Albert Hunt to the Gentlemen of the Legislative Council, 18 September 1865, CAAR 20410 CH287/CP 616/d, PPC 2, Session 24, 23 November 1865, ANZ, Christchurch; *Lyttelton Times*, 24 November 1865, 2; *West Coast Times*, 12 December 1865, 2.

55 'Albert Hunt's claim', *Lyttelton Times*, 7 December 1865, 3.

56 Ihaia Tainui to the Provincial Secretary, 26 December 1863 [translation], ICPS 470 (1)/1863, ANZ, Christchurch.

57 'News from the Grey diggings', *Colonist*, 26 July 1864, 3.

58 Ihaia Tainui to the Provincial Secretary, 15 September 1863, ICPS 570(1)/1863, ANZ, Christchurch.

59 Charlton Howitt to James Wylde, Assistant Engineer, 15 April 1863, ICPS 848/1863, ANZ, Christchurch.

60 Galvin, *Handbook of New Zealand Mines*, p. 149.

61 *Lyttelton Times*, 5 January 1866, 2.
62 *Lyttelton Times*, 5 January 1866, 2; *Press*, 6 January 1866, 3; *Press*, 22 January 1866, 2.
63 'Mr Barff's meeting', *West Coast Times*, 28 March 1866, 2.
64 William Smart to the Superintendent of Canterbury, 26 January 1866, in Alan Oakes, 'Who discovered gold in West Canterbury', Micro-MS-0037, ATL.
65 Provincial Secretary to William Smart, OLPS 160/1866, ANZ, Christchurch.
66 W. A. Taylor, *The Lore and History of the South Island Maori*, Bascands, Christchurch, 1952, pp. 176–77.
67 Martin, in Bradshaw (ed.), *The Diggers' Story*, p. 38. The four parties were Smart and French; a party of Māori diggers; Hunt on his own; and Māori diggers Samuel and Simon, whom Martin had worked with the previous year.
68 Martin, 'A pioneer's reminiscences', p. 49.
69 French later moved to the Buller district, where he ran several accommodation houses and a ferry on a branch of the river. He appears to have given up hope of any recognition or reward from the Canterbury government, for he took no part in Smart's campaign. He never married, and died in Westport in 1887. He was remembered fondly as 'Mickey French', one of the pioneers and 'characters' of the West Coast. 'Obituary: The late Mr. M. P. French', *Grey River Argus*, 8 August 1887, 4; *Westport Times*, 27 June 1868, 4; 12 January 1871, 3; 'Christchurch Militia List', http://christchurchcitylibraries.com/Heritage/Digitised/WarsAndConflicts/NewZealandWars/Militia/.
70 'Obituary: William Smart', *Star*, 12 October 1900, 3; 'Obituary: [Ellen Smart]', *Press*, 12 April 1910, 7.
71 William Smart to Superintendent, 8 August 1871, ICPS 1044/1871, ANZ, Christchurch (it was noted on the letter that there was currently no job vacancy); William Smart to Superintendent, 19 December 1873, ICPS 1946/1873, ANZ, Christchurch; *Press*, 25 January 1876, 4; *Press*, 29 April 1878, 1.
72 *Press*, 12 June 1875, 1.
73 *Press*, 18 February 1870, 4. A company operating under this name was later opened at Lyell Creek, operated by the Zala brothers.
74 *Press*, 12 April 1877, 1.
75 'Smart and Day's new lime kiln', *Press*, 19 November 1877, 3; *Press*, 25 February 1880, 3; *Press*, 17 March 1880, 1; 10 April 1880, 3.
76 During the 1880s and 1890s there were between eight and 12 coal mines operating in the Malvern Hills district. *Appendices to the Journal of the House of Representatives*, 'Reports on the inspection of mines', C-03c to C-05 (1881–1894).
77 *Press*, 4 October 1883, 3; *Star*, 26 November 1883, 3; *Auckland Star*, 29 February 1884, 2.
78 *Press*, 3 September 1881, 4.
79 *Press*, 10 July 1889, 6; *Press*, 17 February 1891, 1.
80 *Otago Daily Times*, 5 September 1890, 5.
81 *Otago Daily Times*, 6 May 1889, 1; *Grey River Argus*, 26 August 1889, 4.
82 *Grey River Argus*, 17 March 1890, 2; *Grey River Argus*, 15 August 1889, 2.
83 *Appendices to the Journal of the House of Representatives*, C-3, 1890, 134.
84 'Inspection of coal mines report', *Appendices to the Journals of the House of Representatives*, C-3B, 1896, 8.
85 'The Hartley Colliery', *Star*, 26 September 1899, 4.
86 'The late Wm Smart, one of Canterbury's oldest settlers', *Weekly Press*, 17 October 1900, 51.
87 Official assignee's bankruptcy files – Christchurch – Smart, William, 22/9/1885, CAMO CH214 101 395; 20/1/1887, CAMO CH214 105 643, ANZ, Christchurch.
88 Rollitt was a travel writer and commentator on New Zealand (and especially Canterbury) life: an astute, lively-minded observer of the world, who described himself as a 'peripatetic philosopher'. *NZ Observer and Free Lance*, 1 June 1895, 8.
89 Frederick Rollitt ('The Warrigal'), 'Kumara to the Greenstone', *Press*, 30 September 1887, 2.
90 William Smart, 'First discoverer of gold in Westland', *Press*, 4 October 1887, 5.
91 Frederick Rollitt ('The Warrigal'), 'Memories of the Grey Valley', *Press*, 19 October 1887, 6.
92 William Smart, 'The discovery of gold on the West Coast: To the editor', *Press*, 14 November 1887, 6.
93 William Smart to the Minister of Mines, and Premier of New Zealand, Richard Seddon, 24 June 1895, ADOL 19382 MD1/180 1895/1002-1895 1100, ANZ, Wellington.
94 Galvin, *Handbook of New Zealand Mines*, pp. 118, 116.

# Weights and measures

Troy ounce (oz t): an old unit of imperial measure, still used for weighing precious metals.
1 pennyweight (dwt) = 24 grains = 1.55 grams
20 pennyweight = 1 troy ounce (oz t) = about 1.097 standard cunces (avoirdupois) = 31 grams
1 standard ounce = 28.3 grams
16 ounces = 1 pound (lb) = 454 grams
1 hundredweight (cwt) = 112 pounds = 51 kilograms
1 inch = 2.5 centimetres
1 foot = 12 inches = 30 centimetres
1 yard = 3 feet = 900 millimetres = 0.9 of a metre
1 acre = 4047 square metres = 0.4 of a hectare
1 mile = 1.6 kilometres

# Glossary

**alluvial gold:** gold deposited by floods, found on flood plains and in river beds
**auriferous:** gold-bearing
**beachline:** the route along the beach
**bottom:** bedrock
**bush lawyer:** tātaramoa, a New Zealand native plant with hooked thorns that snag clothing and rip skin. The colloquial English name is said to refer to the fact that once this thorny plant becomes attached it will not let go until it has drawn blood.
**claim:** a portion of land that a prospector is entitled to occupy for mining purposes, and for which the government has granted the prospector the exclusive rights of winning gold
**colour, the colours:** gold specks visible in a gold-digger's pan, or in a sample of gravel or quartz
**damper:** an unleavened scone, originally made of flour, salt and water, cooked in the ashes of an open fire or in a camp oven
**drift:** rock debris piled up in heaps by the action of water and/or glaciers
**drive:** a horizontal tunnel
**face, working face:** the vertical surface that miners have exposed – a cliff face, the high side of a terrace or end of a tunnel where the mining work is advancing
**flume** (noun): *see* water-race
**flume** (verb): to divert the waters of a stream by creating a flume or water-race, in order to lay bare auriferous ground
**fresh, freshet:** a rush or sudden flood of water in a river or creek after rain
**go mates:** to enter into a partnership with one or more fellow diggers, to share the labour and any gold found. Sealed with a handshake, it appears to have been regarded as legally binding.
**jump a claim:** to take possession of someone else's claim by dishonesty or force
**lead:** a well-defined bed of gold-bearing gravel or sand (wash-dirt)

**mogie** (various spellings): a colonial variation of the Māori word mōkihi or mōkī, meaning a temporary raft used on lakes, estuaries or rivers. Constructed of bundles of reeds, flax stalks or raupō tied together, it was quick to make on a river bank using materials to hand. See http://www.teara.govt.nz/en/object/6015/mokihi.

**paddock** (noun): an area marked out and systematically excavated for procuring wash-dirt in shallow ground; an excavation where wash-dirt was dumped

**paddock** (verb): to work a claim systematically by means of a series of small shallow pits

**pay dirt:** ground containing ore (gold) in sufficient quantity to be profitably extracted; (fig.) profit or success

**prospector's/prospecting claim:** the first claim in a new locality. The finder was allowed to peg out a larger area than normal, usually double, as a bonus.

**quicksilver cradle:** a wooden box placed in a sloping position and fixed on rockers, used for washing gold-bearing gravel. The gold was caught by quicksilver (mercury) in the lower part of the cradle.

**reef or lode:** a mineral vein in rock

**riffles:** strips of wood, iron or other material used to trap gold in a sluice-box

**shaft:** a vertical excavation

**sink in/sink a shaft:** excavate or bore vertically downwards, usually from the surface to the workable gold

**sluice-box:** a long, sloping trough or similar construction, with grooves (riffles) on the bottom, into which water was directed to separate the gold from gravel or sand

**tailings:** the material or detritus left over after the process of separating gold from sand or gravel

**wash, wash-dirt:** auriferous gravels, sands, etc.

**water-race:** an artificial channel constructed to divert water from a river or creek to wash or sluice the dirt

**whata** (Māori): an elevated platform mainly used for storing food

**worked out:** mined until depleted of minerals or gold

**work up on a face:** to expose a working face of a mining site by excavation or sluicing

## Sources

'Glossary', Mindat.org, http://www.mindat.org/glossary, accessed 8 February 2016.

Heinz, W. F., *Prospecting for Gold*, 5th edn, Pegasus Press, Christchurch, 1975, pp. 76–78.

May, Philip Ross, *The West Coast Gold Rushes*, Pegasus, Christchurch, 1962, pp. 534–36.

Orsman, H. W. (ed.), *Dictionary of New Zealand English*, Oxford University Press, Auckland, 1997.

# Bibliography

### Abbreviations

| | |
|---|---|
| ANZ | Archives New Zealand |
| ATL | Alexander Turnbull Library, National Library of New Zealand, Wellington |
| ICPS | Inward Correspondence of the Provincial Secretary (Canterbury) |

## Unpublished sources

Dobson, Arthur Dudley, 'The opening of the Canterbury West Coast: Address delivered before the Rotary Club on Tuesday 25th July 1922', in Robert Allan (ed.), 'Reminiscences: Early Christchurch and Canterbury, 1922', accessed 2 January 2016 from christchurchcitylibraries.com/DigitalCollection/Archives/Archive210/PDF/Reminiscences-1922-Dobson.pdf.

Howitt, H. C., 'Diary of Mr. Howitt during his late search for gold, 12 September–30 November 1862', qMS-1015, ATL.

——, 'Journal of H. C. Howitt, 1 January–20 June 1863', MS-1038, ATL.

Loder, D. S., 'History of the gold discoveries and rushes of the west coast of Nelson and an estimate of their significance', unpublished thesis, Otago University, 1935, Micro-0276a, ATL.

Martin, William, 'A pioneer's reminiscences', [c. 1913], MS206, Hocken Library, University of Otago.

May, Philip Ross, 'The West Canterbury gold rush to 1865: The analysis of a gold rush in the middle of the nineteenth century', unpublished thesis, Canterbury University College, 1953, Canterbury Museum.

McCaskill, Murray, 'Historical geography of Westland before 1914', unpublished PhD thesis, University of Canterbury, 1960, accessed 7 January 2016 at http://ir.canterbury.ac.nz/handle/10092/4762.

Oakes, Alan, 'Who first found gold in West Canterbury?', 1856, Micro-MS-0037, ATL.

Outwards Letters [Assistant Secretary of Native Affairs], 23 February 1858–8 July 1863, MA-Collingwood 2 Box 1, ANZ Wellington.

Revell, J. C., 'Journal of the Grey Store, 30 May to 30 June 1864', ICPS 1998/1864, ANZ, Christchurch.

——, 'Journal of labor, events and occurrences at the Prov. Govt. depot, West Coast, 1 July to 31 July 1864', ICPS 2141/1864, ANZ, Christchurch.

——, 'Report on West Coast business, February 16 1864', ICPS 717/1864; 718/1864, ANZ, Christchurch.

Revell, W. H., 'Correspondence relative to claims made by Albert Hunt and others for the reward

offered for the discovery of a payable gold field in the province of Canterbury', 11 August 1865, Canterbury Provincial Council, Papers and Returns, CAAR CH287 20410 Box CP 618 PPC 9, ANZ, Christchurch.
——, 'Grey depot journal, May 1864', ICPS 1894(1)/1864, ANZ, Christchurch.
——, 'Journal of the Grey River Store, 23 February 1864', ICPS 590/1864, ANZ, Christchurch.
Smart, William, 'Diary of a prospecting tour to the West Coast, from 1862 to 1865', 1865, CAAR CH287 19936 ICPS 790(1)/1865, ANZ, Christchurch.
——, 'The Westland Goldfields, from the diary of Wm Smart, one of the first discoverers of payable gold in that part of N. Zealand', 1890, Microfilm 64/310, Macmillan Brown Library, University of Canterbury, Christchurch.

## Published sources

Acland, L. D. G., *Early Canterbury Runs*, revised edn, Whitcombe & Tombs, Christchurch, 1946.
Bradshaw, Julia (ed.), *The Diggers' Story: Accounts of the West Coast gold rushes* (An annotated and illustrated new edition of the classic publication, compiled by Carl Pfaff and published by the West Coasters' Association in 1914), Canterbury University Press, Christchurch, 2014.
Brailsford, B., *Greenstone Trails: The Maori search for pounamu*, Reed, Wellington, 1984.
Brown, Margaret, 'Christmas 1862 at the mouth of the Buller River', *Nelson Historical Society Journal*, 3, no. 4 (September 1978), accessed 20 January 2016 at http://nzetc.victoria.ac.nz/tm/scholarly/tei-NHSJ03_04-t1-body1-d2-d3.html.
Brunner, Thomas, *The Great Journey: An expedition to explore the interior of the Middle Island of New Zealand, 1846–48*, Pegasus, Christchurch, 1952.
——, *Journal of an Expedition to the Interior of the Middle Island of New Zealand*, William Clowes & Sons, London, 1850.
'Canterbury Land Index', 1853–54, accessed 20 January 2016 at http://www.nzpictures.co.nz/canterburylandindex1853-1864.pdf..
*Cyclopedia of New Zealand: Wellington Provincial District*, The Cyclopedia Company, Wellington, 1897.
Dobson, Arthur Dudley, *The Reminiscences of Arthur Dudley Dobson, Engineer, 1841–1934*, Whitcombe & Tombs, Auckland, 1930.
Galvin, Patrick, *Handbook of New Zealand Mines*, New Zealand Mines Department, Government Printer, Wellington, 1887.
Gardner, W. J. (ed.), *History of Canterbury, Vol. II: General history, 1854–76*, Canterbury Centennial Historical and Literary Committee; Whitcombe & Tombs, Christchurch, 1971.
Haast, J. F., 'Notes on the geology of Canterbury', *Lyttelton Times*, 12 November 1862, 3.
——, *Report of a Topographical and Geological Exploration of the Western Districts of the Nelson Province, New Zealand*, Nelson Provincial Government, printed by C. & J. Elliott, Nelson, 1861.
Harper, H. W., *Letters from New Zealand*, Rees, London, 1914.
Hawes, Carolyn, *Great Expectations: The colonisation of Buller*, Cadsonbury, Christchurch, 2004.
Hawkins, D. N., *Beyond the Waimakariri: A regional history*, Cadsonbury, Christchurch, 1957/2002.
Heaphy, Charles, 'Notes of an expedition to Kawatiri and Araura on the western coast of the Middle Island, performed by Messrs Heaphy and Brunner', *Nelson Examiner & New Zealand Chronicle*, 5 September 1846, 105, to 17 October 1846, 130 [seven weekly instalments].
Heinz, W. F., *Prospecting for Gold*, 5th edn, Pegasus, Christchurch, 1975.
Hochstetter, Ferdinand, 'Lecture on the geology of the province of Nelson', *Nelson Examiner & New Zealand Chronicle*, 1 October 1859, 1.
Latham, Darrell, *The Golden Reefs: An account of the great days of quartz-mining at Reefton, Waiuta and the Lyell*, 2nd edn, Nikau Press, Nelson, 1992.
Loughnan, R. A., *The First Gold Discoveries in New Zealand*, John Mackay, Government Printer, Wellington, 1906.
Low, Hilary, *Pushing his Luck: Report of the expedition and death of Henry Whitcombe, by Jakob Lauper*, Canterbury University Press, Christchurch, 2010.
Mackay, James Jr, 'Notes of a journey from Pakawau, Massacre Bay to Mawera or Grey River on the West Coast, February and March 1857', *Nelson Examiner & New Zealand Chronicle*, 26 August 1857, 3.
'Māori and gold', accessed 20 January 2016 at http://www.theprow.org.nz/maori/maori-and-gold/#.

Martin, Wayne, *Murder on the Maungatapu: A narrative history of the Burgess Gang and their greatest crime*, Canterbury University Press, Christchurch, 2016.
Martin, William, 'A pioneer's birthday story', in Julia Bradshaw (ed.), *The Diggers' Story: Accounts of the West Coast gold rushes* (An annotated and illustrated new edition of the classic publication, compiled by Carl Pfaff and published by the West Coasters' Association in 1914), Canterbury University Press, Christchurch, 2014, pp. 36–45.
May, Philip Ross, *The West Coast Gold Rushes*, Pegasus, Christchurch, 1962.
McCaskill, Murray, 'The Poutini coast: a geography of Maori settlement in Westland', *New Zealand Geographer*, 10, no. 2 (1954), 134–50.
McKinnon, Malcolm (ed.), *New Zealand Historical Atlas*, David Bateman, Auckland, 1997.
Mitchell, Hilary & John Mitchell, *Te Tau Ihu o te Waka: A history of Māori of Nelson and Marlborough, Vol. 2: Te Ara Hou: The new society*, Huia Publishers, Wellington / Wakatū Incorporation, Nelson, 2007.
——, *Te Tau Ihu o te Waka: A history of Māori of Nelson and Marlborough, Vol. 4: Ngā whānau rangatira o Ngāti Tama me Te Ātiawa: The chiefly families of Ngāti Tama and Te Ātiawa*, Huia Publishers, Wellington / Wakatū Incorporation, Nelson, 2014.
Money, Charles, *Knocking about in New Zealand*, Capper Press, Christchurch 1871/1972.
Newport, J. N. W., *Collingwood: A history of the area from earliest days to 1912*, Caxton, Christchurch, 1971.
Oakes, John P., 'Discovery of gold on the West Coast: To the editor of the *Lyttelton Times*', *Lyttelton Times*, 21 March 1865, 5.
Orsman, H. W. (ed.), *Dictionary of New Zealand English*, Oxford University Press, Auckland, 1997.
Pascoe, John, 'The Maori and the mountains: Part 2, Across the Main Divide', *Te Ao Hou*, 12 (1955), 26–27, accessed 10 January 2016 at http://teaohou.natlib.govt.nz/journals/teaohou/issue/Mao12TeA/c19.html.
Phillips, Jock, *A Man's Country? The image of the Pakeha male, a history*, revised edn, Penguin, Auckland, 1996.
Pickering, Mark, 'The colours: The search for payable gold, 1857 to 1864', 2011, accessed 10 January 2016 at http://www.bestwalks.kiwi.nz/uploads/4/9/3/3/49336433/the_colours_new_design.pdf.
Preshaw, George Ogilvy, *Banking under Difficulties, or Life on the Gold Fields of Victoria, New South Wales & New Zealand, by a Bank Official*, Kiwi Publishers, Christchurch, 1888/1997.
Rochfort, John, 'A brief account of an expedition to the West Coast, December 1859', *Nelson Examiner & New Zealand Chronicle*, 24 December 1859, 3.
Rollitt, Frederick ('The Warrigal'), 'Kumara to the Greenstone', *Press*, 30 September 1887, 2.
——, 'Memories of the Grey Valley', *Press*, 19 October 1887, 6.
Salmon, J. H. M., *History of Goldmining in New Zealand*, Government Printer, Wellington, 1963.
Saunders, Alfred, *Tales of a Pioneer: Episodes in the life of Alfred Saunders, selected and arranged by his two youngest daughters*, Isitt-Wildey, Christchurch, 1927.
Sherrin, R. A. A., 'The West Coast: Mr. Sherrin's journal', *Press*, 13, 15, 17, 19, 25, 26 December 1863.
Skinner, H. D., 'Maori life on the Poutini coast, together with some traditions of the natives', *Journal of the Polynesian Society*, 21, no. 84 (1912), 141–51.
Smart, William, 'The discovery of gold on the West Coast', *Press*, 14 November 1887, 6.
——, 'First discoverer of gold in Westland', *Press*, 4 October 1887, 5.
Spencer, W. E., *A History of the Buller District*, masters thesis, University of Otago, 1962; Robert T. Porter, Westport, 1986.
Taylor, W. A., *The Lore and History of the South Island Maori*, Bascands, Christchurch, 1952.
*Te Ara – The Encylopedia of New Zealand*, www.teara.govt.nz.
Waitangi Tribunal, *Ngāi Tahu Report 1991, WAI 27: 13 Arahura*, accessed 4 January 2016 at http://www.justice.govt.nz/tribunals/waitangi-tribunal/Reports/wai0027%201991%20Report/doc_014.
Waite, Reuben, *Narrative of the Discovery of the West Coast Gold-Fields*, W. H. L. Leech (ed.), J. Hounsell, Nelson, 1869.
Ward, Alan Dudley, *A Report on the Historical Evidence: The Ngai Tahu Claim, WAI/27 T1*, 1989, accessed 11 February 2016 at http://repository.digitalnz.org/system/uploads/record/attachment/579/_a_report_on_the_historical_evidence__the_ngai_tahu_.pdf.
Wright, Les, *Greenstone Gold: The first West Coast goldfield, 1864–1939*, Department of Conservation, [Hokitika, 198?].

## Newspapers and periodicals

*Appendices to the Journals of the House of Representatives*
*Canterbury Standard* (Christchurch)
*Colonist* (Nelson)
*Grey River Argus* (Greymouth)
*Lyttelton Times*
*Nelson Evening Mail*
*Nelson Examiner & New Zealand Chronicle*
*Otago Daily Times*
*Otago Witness*
*Press* (Christchurch)
*Star* (Christchurch)
*West Coast Times* (Hokitika)

# Index

Page numbers in **bold** refer to illustrations and information in captions.